Co-Clustering

# Co-Clustering

*Models, Algorithms and Applications*

Gérard Govaert
Mohamed Nadif

*Series Editor*
*Francis Castanié*

iSTE

WILEY

First published 2014 in Great Britain and the United States by ISTE Ltd and John Wiley & Sons, Inc.



ISTE Ltd
27-37 St George's Road
London SW19 4EU
UK

www.iste.co.uk

John Wiley & Sons, Inc.
111 River Street
Hoboken, NJ 07030
USA

www.wiley.com



Library of Congress Control Number: 2013950131

British Library Cataloguing-in-Publication Data
A CIP record for this book is available from the British Library
ISBN: 978-1-84821-473-6

# Table of Contents

# Acknowledgment

This research was supported by the CLasSel ANR project ANR-08-EMER-002.

# Introduction

Many of the data sets encountered in statistics are two dimensional and can be represented by a rectangular numeric table, that is an $n$ by $d$ data matrix $\mathbf{x} = (x_{ij})$ defined on two sets $I$ and $J$, sometimes referred to as two-way or two-mode data. For instance, $I$ may be a set of individuals (observations, cases, objects and persons) and $J$ may be a set of variables (measurements, attributes and features). The data matrix then collects the values taken by all the variables for each individual. These data may be represented either as a table of individuals–variables as in the case of continuous variables, or as a frequency table or contingency table as in the case of categorical variables. In the following we examine a number of types of data on which co-clustering can be performed.

## I.1. Types and representation of data

The type of a variable is determined by the set of possible values that the variable can take. In the following, we briefly review each type.

### I.1.1. *Binary data*

Binary variables are widely used in statistics. Examples include presence–absence data in ecology, black and white pixels in image processing and the data obtained when recoding a table of qualitative variables. Data take the form of a sample x $= (\boldsymbol{x}_1, \ldots, \boldsymbol{x}_n)$ where $\boldsymbol{x}_i$ is a vector $(x_{i1}, \ldots, x_{id})$ of values $x_{ij}$ belonging to the set $\{0, 1\}$. For example, the data might correspond to a set of 10 squares of woodland in which the presence (1) or absence (0) of two types of butterflies P1 and P2 was observed. Figure I.1 illustrates three alternative ways of presenting these data.

| Square | P1 | P2 |
|---|---|---|
| 1 | 1 | 1 |
| 2 | 1 | 0 |
| 3 | 1 | 1 |
| 4 | 0 | 1 |
| 5 | 0 | 1 |
| 6 | 1 | 1 |
| 7 | 1 | 0 |
| 8 | 0 | 0 |

| State | Frequency |
|---|---|
| 00 | 1 |
| 01 | 2 |
| 10 | 2 |
| 11 | 3 |

| P1 \ P2 | 1 | 0 |
|---|---|---|
| 1 | $n_{11} = 3$ | $n_{10} = 2$ |
| 0 | $n_{01} = 2$ | $n_{00} = 1$ |

**Figure I.1.** *Example of binary data*

Binary data have been treated in clustering with a large number of distances, most of which are defined using the values $n_{11}$, $n_{10}$, $n_{01}$ and $n_{00}$ of the table crossing the two variables. For example, the distances between two binary vectors $i$ and $i'$ measured using "Jaccard's index" and the "agreement coefficient" can be written, respectively

$$\mathrm{d}(\boldsymbol{x}_i, \boldsymbol{x}_{i'}) = \frac{n_{11}}{n_{00} + n_{10} + n_{01}} \quad \text{and} \quad \mathrm{d}(\boldsymbol{x}_i, \boldsymbol{x}_{i'}) = n_{11} + n_{00}.$$

### I.1.2. *Categorical data*

Categorical variables, sometimes known as qualitative variables or factors, are a generalization of binary data to situations where there are more than two possible values.

Here, each variable may take an arbitrary finite set of values, usually referred to as *categories*, *modalities* or *levels*. Like binary data, categorical data may be represented in different ways: as a table of individuals–variables of dimension $(n, d)$, as a frequency vector for the different possible states, as a contingency table with $d$ dimensions linking the categories or as a *complete disjunctive table* where categories are represented by their indicators. In this last form of representation, which we will use here, the data are composed of a sample $(\boldsymbol{x}_1, \ldots, \boldsymbol{x}_n)$, where $\boldsymbol{x}_i = (x_i^{jh}; j = 1, \ldots, d; h = 1, \ldots, m_j)$, with

$$\begin{cases} x_{ij}^h = 1 & \text{if } i \text{ takes the modality } h \text{ for the variable } j \\ x_{ij}^h = 0 & \text{otherwise,} \end{cases}$$

where $m_j$ denotes the number of modalities of the variable $j$. In Figure I.2, a data matrix is shown which consists of a set of eight individuals described by three categorical variables A, B and C and its associated complete disjunctive table.

| | A | B | C |
|---|---|---|---|
| 1 | 1 | 1 | 1 |
| 2 | 2 | 1 | 1 |
| 3 | 1 | 3 | 1 |
| 4 | 2 | 2 | 1 |
| 5 | 2 | 2 | 2 |
| 6 | 1 | 3 | 2 |
| 7 | 1 | 1 | 3 |
| 8 | 1 | 1 | 3 |

| | A1 | A2 | B1 | B2 | B3 | C1 | C2 | C3 |
|---|---|---|---|---|---|---|---|---|
| 1 | 1 | 0 | 1 | 0 | 0 | 1 | 0 | 0 |
| 2 | 0 | 1 | 1 | 0 | 0 | 1 | 0 | 0 |
| 3 | 1 | 0 | 0 | 0 | 1 | 1 | 0 | 0 |
| 4 | 0 | 1 | 0 | 1 | 0 | 1 | 0 | 0 |
| 5 | 0 | 1 | 0 | 1 | 0 | 0 | 1 | 0 |
| 6 | 1 | 0 | 0 | 0 | 1 | 0 | 1 | 0 |
| 7 | 1 | 0 | 1 | 0 | 0 | 0 | 0 | 1 |
| 8 | 1 | 0 | 1 | 0 | 0 | 0 | 0 | 1 |

**Figure I.2.** *Example of categorical data (left) and its associated complete disjunctive table (right)*

### I.1.3. *Continuous data*

Continuous data are undoubtedly the most current type of data and can be found in all areas. The structure takes the form of a relational table where the $d$ columns are continuous variables and $\boldsymbol{x}_{ij} \in \mathbb{R}$. They can be positive or negative with different units and variabilities. The measurement unit used can affect the results of different methods of data analysis and a normalization or transformation is often necessary. For instance, a variable can be normalized by scaling its values so that they lie within a specified range, such as $[0, 1]$. This aim can be achieved by the min-max normalization defined by

$$\frac{x_{ij} - \mathrm{min}_j}{\mathrm{max}_j - \mathrm{min}_j},$$

where $\mathrm{min}_j$ and $\mathrm{max}_j$ are, respectively, the lowest and the highest values taken by the variable $j$. The logarithmic transformation is also commonly used to pre-process data. These two transformations are frequently used with microarray data sets in order to overcome problems of inaccuracy of measurement or to provide values that are more easily interpretable. Other transformation techniques exist and are commonly used. We can cite, for instance, the $z$-score normalization defined by

$$\frac{x_{ij} - \mu_j}{\sigma_j},$$

where $\mu_j$ and $\sigma_j$ are, respectively, the mean and the standard deviation of the variable $j$. Sometimes, and in order to reduce the effect of outliers, a variation of this $z$-score normalization consists of replacing $\sigma_j$ by $s_j$, the mean absolute deviation of $j$. Different ways to normalize the data also exist. The user should pay special attention to this step as it is essential for obtaining meaningful results.

Besides, most authors distinguish two types of analysis: Tryon and Bailey [TRY 70] suggest "0-Analysis" for the study of objects and "V-Analysis" for the study of variables. According to them, the earliest works relate to the analysis of objects, which is the classification (taxonomy). The first work on the analysis of the variables, from Pearson and Spearman, is the factor analysis. In other domains, these two types of analysis are called "P-technique" and "Q-technique".

In the data previously described, both sets (individuals and variables) show a strong asymmetry, however in some situations the two sets play a similar role and can be interchanged. The contingency table studied in the next section is the most common example of this type of data.

### I.1.4. *Contingency table*

There are many situations where we try to study the association between two categorical variables. A two-way contingency table is a method for summarizing the two variables. We can remark that this definition can be easily extended to more categorical variables. With data of this kind, the cells, formed by the cross-tabulation of two categorical variables, $I$ having $n$ categories and $J$ having $d$ categories, contain the frequency counts of the individuals belonging to these cells. Contingency tables of this sort can be found in many distinctive applications. An important example is information retrieval and document clustering, where $I$ may correspond to a collection of documents and $J$ to a set of words, the frequency denotes the number of occurrences of a word in a document. It is also noteworthy that the definition of the contingency table can also be extended to tables where every entry expresses a quantity of the same matter, in such a way that all of the entries can be meaningfully summed up to a number expressing the total amount of matter in the data. Examples of such data are trade tables showing the money transferred from country $i$ to country $j$ during a specified period. We now specify the notation that will be used to study the contingency table.

Let $\mathbf{x} = (x_{ij}, i = 1, \ldots, n; j = 1, \ldots, d)$ be a two-way contingency table associated with two categorical random variables that take values in sets $I = \{1, \ldots, n\}$ and $J = \{1, \ldots, d\}$. The entries $x_{ij}$ are co-occurrences of row and column categories, each of which counts the number of entities that fall simultaneously into the corresponding row and column categories. The sum of frequencies of row and column categories, usually called marginals, are denoted by $x_{i.}$ and $x_{.j}$ and defined by $x_{i.} = \sum_j x_{ij}$, $x_{.j} = \sum_i x_{ij}$ and $x_{..} = \sum_{i,j} x_{ij}$. Here, we use the usual dot notation to express the sum with respect to the suffix replaced by a dot. Let $P_{IJ} = (p_{ij})$ denote the sample joint probability distribution. It is a matrix of size $n \times d$ defined by $p_{ij} = \frac{x_{ij}}{N}$ where $N = x_{..}$. The sample marginal probability distributions are defined by $p_{i.} = \sum_j p_{ij}$ and $p_{.j} = \sum_i p_{ij}$. The sample joint probability distribution $p_{ij}$ can be considered as estimators of the probabilities $\xi_{ij}$ that the two categorical random variables occur in the cell in row $i$ and column $j$. Table I.1 presents the form of the contingency table and of the corresponding sample joint distribution.

| | 1 | … $j$ | … $d$ | |
|---|---|---|---|---|
| 1 | $x_{1j}$ | … $x_{1j}$ | … $x_{1d}$ | $x_{1.}$ |
| | | ⋮ | … | |
| $i$ | $x_{i1}$ | … $x_{ij}$ | … $x_{id}$ | $x_{i.}$ |
| | | ⋮ | … | |
| $n$ | $x_{n1}$ | … $x_{nj}$ | … $x_{nd}$ | $x_{n.}$ |
| | $x_{.1}$ | … $x_{.j}$ | … $x_{.d}$ | $N$ |

| | 1 | … $j$ | … $d$ | |
|---|---|---|---|---|
| 1 | $p_{1j}$ | … $p_{1j}$ | … $p_{1d}$ | $p_{1.}$ |
| | | ⋮ | … | |
| $i$ | $p_{i1}$ | … $p_{ij}$ | … $p_{id}$ | $p_{i.}$ |
| | | ⋮ | … | |
| $n$ | $p_{n1}$ | … $p_{nj}$ | … $p_{nd}$ | $p_{n.}$ |
| | $p_{.1}$ | … $p_{.j}$ | … $p_{.d}$ | 1 |

**Table I.1.** *Contingency table and sample joint distribution*

Sometimes, and specifically in document clustering when the rows are documents and the columns are words, some transformations of data are necessary. For instance, the co-occurrences can be replaced by the tf-idf statistics

[JON 72]. Different variants are proposed and commonly used in information retrieval and text mining.

### I.1.5. *Data representations*

Different representations can be associated with the types of data described in the previous section.

*Geometrical representation:* for the continuous data, a classical geometrical representation consists of regarding these data as $n$ points in $d$ dimensions. In a dual way, a second and less familiar geometrical representation consists of regarding the data as $d$ points in $n$ dimensions. The classical methods, such as principal component analysis and $k$-means algorithm, used such representations extensively. Correspondence analysis [BEN 73b] uses similar geometrical representations to the contingency table.

*Bipartite graph:* in all situations, it is possible to associate the data matrix to a *bipartite graph* whose vertices are the elements of the union $I \cup J$ of sets $I$ and $J$. For individuals×variables table and the contingency table, the edges of the graph are the set of pairs $\{(i, j), i \in I, j \in J\}$ weighted by corresponding entries $x_{ij}$ in the data matrix. For binary data, the edges of the graph are the set of pairs $(i, j)$ such that $x_{ij} = 1$ (see, for instance, Figure I.3). This representation is frequently considered in the graph community such as in Web 2.0 tagging data and social networks.

The methods we are interested in next are clustering methods and, specifically, the simultaneous clustering of $I$ and $J$. To this end, we will review the motivation of simultaneous analysis and then introduce co-clustering.

| | B1 | B2 | B3 | B4 | B5 | B6 | B7 |
|---|---|---|---|---|---|---|---|
| A1 | 1 | 1 | 1 | 0 | 0 | 1 | 1 |
| A2 | 1 | 1 | 0 | 0 | 0 | 1 | 1 |
| A3 | 0 | 0 | 0 | 1 | 1 | 1 | 1 |
| A4 | 0 | 0 | 0 | 1 | 0 | 1 | 1 |

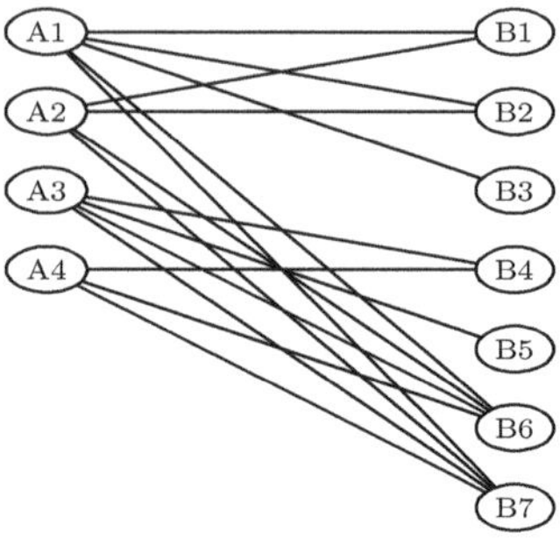


**Figure I.3.** *Binary data and its associated bipartite graph*

## I.2. Simultaneous analysis

### I.2.1. *Data analysis*

Given a data matrix, the objective of data analysis can be viewed as the simultaneous analysis of the two sets $I$ and $J$ to identify underlying structures that may exist between these two sets. Different approaches such as exploratory analysis (graphical representation or numerical summary) or dimension reduction have been used. Principal component analysis and correspondence analysis are examples of such methods. This last method given by Benzecri [BEN 73b] is one of the best known methods that performs analysis *simultaneously* on both sets $I$ and $J$. The data table must be a contingency table or at least it must have similar properties. The properties of this approach, especially transition formulas, allow us to exchange the results on the sets $I$ and $J$. These properties help us to define a set of barycentric relations, justifying a simultaneous representation of $I$ and $J$ and allowing us to simultaneously visualize the proximity among the elements of $I$, the elements of $J$ and the elements of $I$ and $J$. Finally let us quote the *unfolding method* of [COO 50] for which the objective is to represent rank preference data on a line or a plan. Each individual is represented by an ideal point such that the relation of order among the variables, defined by the distances between the ideal point and the various variables, is closest to the order given in the initial data.

Other methods relate to direct processing of the data matrix. For instance, seriation methods amount to finding a permutation of rows associated with a permutation of columns, leading to a reshaped data matrix with a maximum density of high cell values along the diagonal, in addition to low value areas in the upper and lower parts. Such approaches have been used, for instance, in archaeology, phytosociology, geography and production management. Caraux [CAR 84] proposed a criterion based on an objective function with quadratic costs and Bertin [BER 80] proposed a manual heuristics based on visual densification. Factorial methods such as correspondence analysis can also be used. Note that when correspondence analysis gives rise to a U-shaped effect (Guttman effect) on the first two axes of the factorial representation, there exists a latent order within the rows and the columns leading to diagonal band reshaping, which corresponds to the order of the projections along the first axis of the rows and columns.

This book is devoted to another group of methods of simultaneous analysis of two sets by using the notion of clustering. With a two-way or two-mode data set, clustering algorithms are often applied to just one mode of the data matrix, which can be done in a hierarchical or non-hierarchical way. Among the non-hierarchical methods, $k$-means clustering [FOR 65, MAC 67, HAR 75b] is one of the most popular methods. Contrary to this approach, there is a relatively new form of clustering that analyzes the two sets simultaneously. These methods, called direct clustering, cross-clustering, simultaneous clustering, co-clustering, biclustering, two-way clustering, two-mode clustering or two-side clustering, have developed considerably in recent times.

### I.2.2. *Co-clustering*

A large number of co-clustering algorithms have been proposed to date. One of the earliest and most cited biclustering formulations, known as block clustering, was proposed by Hartigan [HAR 72, HAR 75a]. He sought to organize the data table using structures that may be, for example, defined from classifications on each of the two sets. This kind of method is sometimes known as direct clustering. Older works can also be cited. For instance, this problem was first described formally by Good [GOO 65] who proposed a technique for the simultaneous clustering of objects and variables. Fisher [FIS 69] posed the problem of the simultaneous search for clustering on the row and column dimensions of a data matrix in a metric way. He defined a criterion for optimization, but offered no method to solve this problem. Tryon and Bailey [TRY 70] first clustered the set of variables using the correlation matrix and then, using a distance measure across the clusters of variables, clustered the set of individuals. Dubin and Champoux [DUB 70] proposed a method that combines the variables into types, and associates each individual with the types of variables forming a classification of individuals. More often, the authors discussed the classification of individuals, describing at length the choice of a measure of similarity and merely mentioned the possibility of a classification of variables without dwelling on how to get there. Anderberg [AND 73] identified the choice between $I$ and $J$ among the list of the problems of classification. He considered it reasonable to classify variables as individuals. He even suggested an iterative approach in which the classification is done alternately on the individuals and the variables until the classifications on both sets are mutually "harmonious", believing that such research simultaneously offers "considerable potential to increase the effectiveness of automatic classification".

In the case of contingency tables and using the chi-squared statistic $\chi^2$ as measure of information, Govaert [GOV 77] developed an algorithm, called CROKI2 in the following to simultaneously search for partitions of each set, minimizing the loss of information due to regrouping the two sets into classes. Extending this approach to binary, continuous and categorical data, Govaert [GOV 83] proposed the CROBIN, CROEUC and CROMUL algorithms. Bock [BOC 79] showed the interest of the simultaneous classification and gave several examples of problems for which a good solution is provided by a simultaneous classification. Since that time, this area has developed considerably and particularly in text mining (see, for instance, [DHI 03]). An extensive overview of two-mode clustering methods can be found in [VAN 04] and, in the context of biological data analysis, in [MAD 04].

### I.2.3. *Applications*

Throughout the last four decades, biclustering has been used in many diverse areas. In this section, the aim is not to give an exhaustive list, but to briefly describe some applications and to provide some information about the situation of their use.

*Text mining:* in information retrieval systems, the model commonly used to represent the data is the bag-of-words or vector space model [SAL 83]. A set of words is chosen from the set of all words in all documents. Each document is a vector in the feature space formed by these words. The vector entries can be frequencies or some other measures. Thus, the entire document collection may be represented by a word-by-document matrix whose rows correspond to words and columns to documents. Generally, each document contains only a small number of words and hence, the data matrix is very sparse. Since the data dimension may be huge, a lower dimensional representation is imperative for efficient manipulation and co-clustering is a reference tool to summarize the data. We can cite the works of co-clustering documents and words using bipartite spectral graph partitioning in [DHI 01] or information-theoretic in [DHI 03].

*Web mining:* web clustering is an approach for aggregating Web objects into various groups according to underlying relationships among them. The classical clustering algorithms are mainly used only on one dimension of the data, i.e. on user dimension or on page dimension, rather than taking into account the correlation among users and pages. Because the ultimate goal is to extract subsets of user sessions and Web pageviews to construct a variety of co-clusters, Web co-clustering is an effective means to address this challenge. Xu *et al.* [XU 10] proposed a procedure based on a bipartite spectral projection approach. To categorize the Web pages, Charrad *et al.* [CHA 09] used the CROKI2 co-clustering algorithm [GOV 77].

*Bioinformatics:* gene expression profiling has been established over the last two decades as a standard technique for obtaining a molecular fingerprint of tissues or cells in different biological conditions. The technology of microarrays enables us to measure mRNA levels for thousands of genes simultaneously [DER 96]. Given a set of gene expression profiles, organized as a large data matrix illustrating the expression levels of genes (rows of the matrix) under different samples, such as tissues or experimental conditions (columns of the matrix), a common aim is to identify subsets of gene whose expression levels exhibit a coherent pattern under a subset of conditions. Several works concern this topic and we will give a non-exhaustive list in section I.4.4. The interested reader can find a structured overview of the different algorithms used in [MAD 04] and [TAN 05].

*Marketing:* the objective of recommender systems is to predict individual choices and preferences based on observed preference behavior. Collaborative filtering is a technique used by some recommender systems, it consists of matching data of one user with data of similar users, based on purchasing and browsing patterns [GOL 92, HOF 99b]. Thus, it allows merchants to provide customers with future purchase recommendations. In this situation, biclustering is a good solution. For instance, for a recommender system in the movie domain, because data are always sparse, much more accurate predictions can be made by grouping people into clusters with similar movie preferences and grouping movies into clusters that tend to be liked by the same people. Current collaborative filtering techniques such as correlation and singular value decomposition methods are commonly used but are very expensive and can only be deployed in static off-line settings. To address these problems, in [GEO 05], the author proposes a collaborative filtering approach based on a weighted co-clustering algorithm [BAN 07] that involves the simultaneous clustering of users and items.

*Ecology:* in this domain the data often take the form of contingency tables defined by the cover-abundance scores of a set of species in a set of sample units (quadrat, lake and county). Quite often, retaining only the information of presence or absence, the data take the form of binary data [POD 91]. It can also be continuous data. Bock [BOC 79] cited an agricultural research institute that focused on the performance of a set of varieties of fruits planted in different regions. The yields calculated for each variety and each region define continuous data.

*Group technology:* group technology is an approach that was widely used in many industries, including the design of job-shops and flexible manufacturing systems. Group technology is also very important for designing cellular manufacturing systems. In this situation, $I$ is a set of $n$ parts, $J$ is a set of $d$ machines and $x_{ij}$ is the processing time of part $i$ using the $j$th machine. Cellular manufacturing involves processing a collection of similar parts (part families) on a dedicated cluster (or cell) of machines or manufacturing processes. This problem can be addressed by co-clustering. In [GAR 86], the authors proposed a cross-decomposition algorithm called group production management (GPM) and showed that it outperforms the bond energy algorithm (BEA) introduced by McCormick *et al.* [MCC 72]. Note that in [MAR 87], the same objective was studied.

*Archeology:* Lerredde and Perin [LER 80] worked on a set of Merovingian buckle plates for which the presence or absence of a selection of criteria for manufacturing techniques, shape and decoration was observed. The problem was to structure the data by a series of permutations of rows and columns to show links between criteria and plates. The objective was to establish a typology of plates and criteria (biclustering problem) as well as to highlight a temporal evolution in manufacturing techniques (seriation problem).

*Computer science:* to study the program behavior via reference string analysis, the CROKI2 co-clustering algorithm given by Govaert [GOV 77] was used in [SCH 77a], [SCH 77b] and [SCH 83]. Besides, the BEA was used for many applications such as imaging, ordering and related engineering problems and database attribute fragmentation [HOF 75, NAV 84]. In spatial databases, the cost of a spatial join can be very high due to the large size of spatial objects and the computation spatial operationally intensive. In [JIT 01], the authors proposed a variant of BEA to reduce the I/O cost of spatial-join processing. Recently, Bisson and Grimal [BIS 12] proposed a parallelizable approach for the co-clustering of multiview data sets.

## I.3. Notation

In this section, we introduce notations that will be used throughout the book.

– The sums and the products relating to rows, columns, row clusters and column clusters will be subscripted, respectively, by the letters $i$, $j$, $k$ and $\ell$, without indicating the limits of variation which will be implicit. So, the sums $\sum_i$, $\sum_j$, $\sum_k$ and $\sum_\ell$ stand, respectively, for $\sum_{i=1}^n$, $\sum_{j=1}^d$, $\sum_{k=1}^g$, and $\sum_{\ell=1}^m$.

– In the same way, an $n \times d$ matrix a with elements $a_{ij}$ for $i = 1, \ldots, n$ and $j = 1, \ldots, d$ will simply be identified $\mathrm{a} = (a_{ij})$.

– Data will be denoted by an $n$ by $d$ data matrix $\mathrm{x} = (x_{ij}, i \in I = \{1, \ldots, n\}; j \in J = \{1, \ldots, d\})$.

– A partition of $I$ into $g$ clusters will be represented by the classification matrix $\mathbf{z} = (z_{ik}, i = 1, \ldots, n, k = 1, \ldots, g)$ where $z_{ik} = 1$ if element $i$ arose from cluster $k$ and $z_{ik} = 0$ otherwise. The $k$th cluster corresponds to the set of elements $i$ such that $z_{ik} = 1$ and $z_{ik'} = 0\ \forall k' \neq k$. Thus, the partition can be represented by a matrix of elements in $\{0, 1\}^g$ satisfying $\sum_k z_{ik} = 1$. For example, the partition made up of the two classes $\{1, 3, 4\}$ and $\{2, 5\}$ is denoted as

$$\mathbf{z} = \begin{pmatrix} z_{11} & z_{12} \\ z_{21} & z_{22} \\ z_{31} & z_{32} \\ z_{41} & z_{52} \\ z_{51} & z_{52} \end{pmatrix} = \begin{pmatrix} 1 & 0 \\ 0 & 1 \\ 1 & 0 \\ 1 & 0 \\ 0 & 1 \end{pmatrix}.$$

Notation $(z_1, \ldots, z_n)$ where $z_i \in \{1, \ldots, g\}$ represents the cluster of $i$ will also be used. Similarly, notation $\mathrm{w} = (w_{j\ell}, j = 1, \ldots, d, \ell = 1, \ldots, d)$ and $(w_1, \ldots, w_d)$ where $w_j \in \{1, \ldots, g\}$ represents the cluster of $j$ will be used for partitions of $J$.

– In the same way, fuzzy partitions of $I$ and $J$ will be represented, respectively, by a matrix $\widetilde{\mathrm{z}}$ of elements in $[0, 1]$

satisfying $\sum_k \widetilde{z}_{ik} = 1$ for all $i = 1, \ldots, n$ and by matrix $\widetilde{\mathrm{w}}$ of elements in $[0,1]$ satisfying $\sum_\ell \widetilde{\mathrm{w}}_{j\ell} = 1$ for all $j = 1, \ldots, d$. With the same example, we can have

$$\tilde{\mathbf{z}} = \begin{pmatrix} \widetilde{z}_{11} & \widetilde{z}_{12} \\ \widetilde{z}_{21} & \widetilde{z}_{22} \\ \widetilde{z}_{31} & \widetilde{z}_{32} \\ \widetilde{z}_{41} & \widetilde{z}_{52} \\ \widetilde{z}_{51} & \widetilde{z}_{52} \end{pmatrix} = \begin{pmatrix} 0.8\ 0.2 \\ 0.1\ 0.9 \\ 0.7\ 0.3 \\ 0.7\ 0.3 \\ 0.4\ 0.6 \end{pmatrix}.$$

– As for the contingency table, we use the usual dot notation to express the sum with respect to the suffix replaced by a dot. For instance, $x_{i.}$, $x_{.j}$ and $x_{..}$ will be used for data matrix x, and $z_{.k}$ and $w_{.\ell}$ will represent the cardinalities of clusters.

– $A^t$ denotes the transpose of matrix $A$.

– The use of the counting measure for discrete data allows us to unify the notation and "pdf" will be used in this book for both the probability density function in the continuous situation and the probability mass function in the discrete situation.

## I.4. Different approaches

### I.4.1. *Two-mode partitioning*

As we have seen in the previous section, a wide variety of procedures have been proposed for finding patterns in data matrices. These procedures differ in the patterns they seek, the types of data they apply to and the assumptions on which they are based. Here, we are concerned only with co-clustering defined by a partition of objects and a partition of variables.

#### I.4.1.1. *Example*

Clustering, which is the process of partitioning a set of objects or a set of variables, is a challenging research field.

All works on data mining devote a great part to clustering (see, for instance, [HAN 12a]). However, related problems with high-dimensionality and sparsity show the limits of clustering and in order to overcome them, two-mode partitioning offers a good solution. We can illustrate the significance of two-mode partitioning with a simple example. In Figure I.4(a), we present a binary data set described by the set of $n = 10$ objects $I$={A,B,C,D,E, F,G,H,I,J} and a set of $d = 7$ binary variables $J = \{1, 2, 3, 4, 5, 6, 7\}$. Figure I.4(b) consists of data reorganized by a partition of $I$ into $g = 3$ clusters a={A,C,H}, b={B,F, J} and c={D,G, I,E}. Figure I.4(c) consists of data reorganized by the same partition of $I$ and a partition of $J$ into $m = 3$ clusters I={1, 4}, II ={3, 5, 7} and III ={2, 6}. Figure I.4(d) clearly reveals an interesting pattern and shows another advantage of co-clustering methods that reduce the initial data matrix x into a simpler data matrix with the same structure. In this example, our initial $(10 \times 7)$ binary data matrix is reduced to a $(g \times m) = (3 \times 3)$ summary binary data matrix. Note that different techniques using clustering algorithms on each set of data matrix can be used to reach the co-clustering objective; however, as we will see, these techniques reveal themselves to be less effective.

#### I.4.1.2. *Clustering toward two-mode clustering*

While the goal is often the simultaneous study of two sets, many researchers have performed two-way clustering by applying algorithms to both sets separately and independently but with a simultaneous analysis of results. We can cite some examples of this type of approaches. In an article discussing the use of data analysis for architectural design, Maroy and Peneau [MAR 72] defined their goal as "the study of the correspondences between object classes and feature classes". For this, they performed a classification to obtain a partition of the individuals and a partition of the characteristics. Then they examined the correspondence between the two classifications with the original table arranged according to an order respecting these partitions. We will return to this concept of ordered arrays later. The parallel use of correspondence analysis also enabled them to study the links between the two classifications. Lerman and Leredde [LER 77] followed the same approach in an application on the characterization of file systems provided by different computer manufacturers. A partitioning method around the nuclei is used to classify the two sets. The intersection of the two partitions obtained is then made in order to allow the interpretation of results. In this study, again, the correspondence analysis is used in conjunction with the classification method. In bioinformatics, Tibshirani *et al.* [TIB 99] illustrated several methods for the two-way visualization of a reordered data matrix based on separately clustering genes and samples using two-way average linkage hierarchical clustering and two-way $k$-means clustering. We can find a similar approach for contingency tables in [CIA 05].

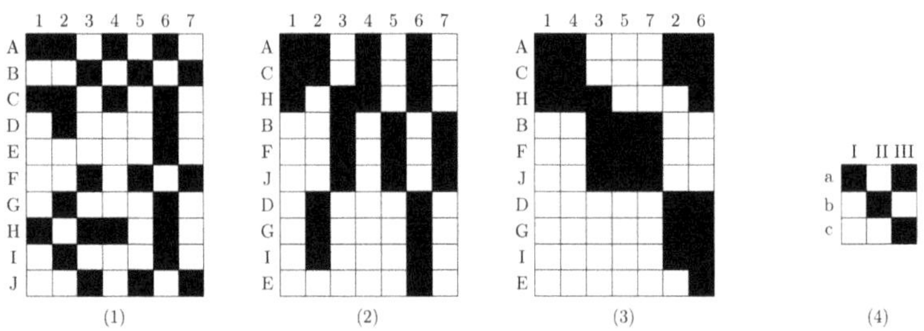


**Figure I.4.** *a) Initial binary data matrix. b) Data matrix reorganized according to a partition of rows. c) Data matrix reorganized according to partitions of rows and columns. d) Summary of this matrix*

As mentioned earlier, a more integrated approach is to classify one of the first sets and then, taking this classification into account, classifying the latter. Using the information bottleneck method introduced by Tishby *et al.* [TIS 99] for finding the best trade-off between accuracy and complexity (compression) when summarizing (e.g. clustering) a random variable, Slonim and Tishby [SLO 00] proposed a two-stage clustering procedure for co-occurrence data: the first stage uses a distribution clustering algorithm to obtain row clusters; in the second stage, these row clusters replace the original rows and a similar procedure is used to obtain column clusters. However, the most interesting situation consists of seeking both partitions simultaneously.

#### I.4.1.3. *Criteria and algorithms*

As for the partitioning clustering situation, the most frequent approach consists of defining a clustering criterion and then finding an algorithm optimizing this criterion. Therefore, the problem is to find the couple of partitions $(\mathrm{z}, \mathrm{w})$, optimizing a function $F(\mathrm{z}, \mathrm{w})$, which expresses the deviation existing between the couple $(\mathrm{z}, \mathrm{w})$ and the initial data $\mathrm{x}$. The form of the criteria depends on the data.

##### I.4.1.3.1. Continuous data

For continuous data, the most frequent criterion used is the least-squares criterion [GOV 83, GOV 95] that can be written as

$$F(\mathbf{z}, \mathbf{w}) = \sum_{i,j,k,\ell} z_{ik} w_{j\ell} (x_{ij} - \overline{x}_{k\ell})^2, \qquad [I.1]$$

where $\overline{x}_{k\ell}$ is the mean of the submatrix defined by the clusters $k$ and $\ell$

$$\overline{x}_{k\ell} = \frac{\sum_{i,j} z_{ik} w_{j\ell} x_{ij}}{z_{.k} w_{.\ell}}.$$

Adding a matrix a $= (a_{k\ell})$ of size $(g \times m)$ where $a_{k\ell}$ is a value associated with each couple of classes $k, \ell$, an extended version of this criterion can be defined in the following way

$$F(\mathbf{z}, \mathbf{w}, \mathbf{a}) = \sum_{i,j,k,\ell} z_{ik} w_{j\ell} (x_{ij} - a_{k\ell})^2. \qquad \text{[I.2]}$$

Two remarks can be made: (1) for fixed partitions z and w, the optimal values $a_{k\ell}$ are the means $\overline{x}_{k\ell}$ and therefore, the optimal partitions for the two criteria are the same partitions. (2) The matrix a, which has the same form as the initial data matrix x (real values), can be viewed as a summary of this matrix. Furthermore, Bock [BOC 79] extended this criterion and developed two variants: a no-interaction model with $a_{k\ell}$ taking this form $a_{k\ell} = \alpha + \beta_k + \gamma_\ell$ and an interaction model with $a_{k\ell} = \alpha + \beta_k + \gamma_\ell + \delta_{k\ell}$. In the first case, it is easy to show that the problem divides into two independent problems of search for partitions on each unit. The usual procedures of search for partitions can therefore be used. The optimal partition pair may be found by applying the one-way sum of squares clustering criterion separately on the rows and the columns. Thus, the usual $k$-means procedure can be used. Introducing the values

$$y_{ij} = x_{ij} - \overline{x}_{i.} - \overline{x}_{.j} + \overline{x}_{..},$$

the second situation is equivalent to criterion [I.1] with new values $y_{ij}$.

Furthermore, there are a variety of methods for matrix approximation using matrix factorization as the principal component analysis for analyzing the rows and columns of a data matrix simultaneously. For example, we can cite the biclustering of gene expression data by non-smooth non-negative matrix factorization [CAR 06]. In fact, when x is positive, even though the co-clustering problem is not the main objective of non-negative matrix factorization (NMF), this approach has attracted many authors for data co-clustering and particularly for document clustering. Given a non-negative matrix x, different algorithms consist of seeking a three-factor decomposition $\mathbf{zaw}^t$ with all factor matrices restricted to be non-negative such as

$$\underset{\mathbf{z}\geq 0,\mathbf{a}\geq 0,\mathbf{w}\geq 0}{\arg\min} ||\mathbf{x} - \mathbf{zaw}^t||^2, \quad [I.3]$$

where $||.||$ is the Frobenius norm.

The matrices z of size $(n \times g)$ and w of size $(d \times m)$ play the roles of row and column memberships and are not necessarily binary. Each value of both matrices z and w corresponds to the degree in which a row or column belongs to a cluster. The matrix a makes it possible to absorb the scales of z, w and x. All proposed algorithms are iterative, which can be differentiated by the update rules of the three matrices due to the chosen optimization method or the supplementary constraints imposed on the three matrices.

The approximation of x can be solved by an iterative alternating least-squares optimization procedure. For instance, the non-negative block value decomposition (NBVD) given by Long *et al.* [LON 05] offers a solution to this problem. Note that the solution to minimizing the criterion [I.3] is not unique. The authors proposed, at the convergence, to normalize each column of the matrix za to have the unit $L^2$ norm. The requirement of normalizing this matrix can be achieved using the new matrix zab where b is a diagonal matrix. The cluster labels of the columns are then deduced from the matrix $\mathrm{b}^{-1}\mathrm{w}^t$ instead of $\mathrm{w}^t$. We can also deduce the label cluster rows by working on the matrix $\mathrm{x}^t$. Note that in NBVD, only the non-negativity of the three matrices is required. In [DIN 06] and [YOO 10], the authors emphasized the importance of the orthogonality constraint, they introduced it on the matrices z and w and proposed, respectively, two variants of orthogonal non-negative matrix tri-factorization called ONM3F in [DIN 06] and ONMTF in [YOO 10]. They can be differentiated by the update rules of the three factors. In a document clustering task, NBVD, ONM3F and ONMTF were shown to work well. Labiod and Nadif [LAB 11a] proposed a co-clustering framework based on NMF formulation. Contrary to previous approaches, they placed the co-clustering aim under the non-negative factorization at the beginning. The key idea is that the latent block structure in a rectangular non-negative data matrix is factorized into two factors rather than three factors, the row-coefficient matrix r and column-coefficient matrix c indicating, respectively, the degree in which a row and a column belong to a cluster. Then, under this framework they developed a variant co-clustering algorithm for non-negative data, which iteratively computes two factors based on two multiplicative update rules. The proposed approach optimizes a relaxed formulation of the double $k$-means criterion in an NMF style, then the optimization procedure looks for the best approximation of the matrix x by the matrix $\mathrm{rr}^t\mathrm{xcc}^t$ with respect to some suitable constraints on the matrices r and c.

I.4.1.3.2. Binary data

In this situation, a natural choice is to search for homogeneous blocks, i.e. blocks with a majority of ones or a majority of zeros. In this case, the objective is to minimize the criterion

$$F(\mathbf{z}, \mathbf{w}) = \sum_{i,j,k,\ell} z_{ik} w_{j\ell} |x_{ij} - a_{k\ell}|, \qquad [I.4]$$

where $a_{k\ell}$ is the modal value of the submatrix defined by the clusters $k$ and $\ell$. This criterion is a direct extension of the well-known maximal predictive classification criterion proposed by Gower [GOW 74].

Additionally, the non-negative matrix factorization approach has been used to identify biclustering in microarray data. Zhang *et al.* [ZHA 10] extended standard NMF to binary matrix factorization (BMF) in order to solve the biclustering problem.

#### I.4.1.3.3. Contingency table

For this type of data, the most common criteria are usually based on the concept of information measures such as the phi-squared coefficient or mutual information (see Chapter 4). The phi-squared coefficient, which can be written as

$$\phi^2(I,J) = \sum_{i,j} \frac{(p_{ij} - p_{i.}p_{.j})^2}{p_{i.}p_{.j}},$$

represents the deviation between the theoretical frequencies $p_{i.}p_{.j}$ that we would have whether $I$ and $J$ were independent and the observed frequencies $p_{ij}$. It usually provides statistical evidence of a significant association, or dependence among the rows and columns of the table. If $I$ and $J$ are independent, the $\phi^2$ will be zero and if there is a strong relationship between $I$ and $J$, the $\phi^2$ will be high. So, a significant phi-squared coefficient indicates a departure from row or column homogeneity and can be used as a measure of the information brought by a contingency table. Various methods [GOO 85] have been proposed for investigating this association. Some of them are graphical approaches and the best known is the correspondence analysis.

Given a couple of partitions $(\mathbf{z}, \mathbf{w})$, a new contingency table $\mathbf{x}^{\mathbf{zw}} = (\mathbf{x}^{\mathbf{zw}}_{k\ell})$ can be defined by regrouping the rows and columns according to the partitions and it can be shown that the phi-squared coefficient $\phi^2(\mathbf{z}, \mathbf{w})$ associated with this new contingency table verifies

$$\phi^2(I,J) > \phi^2(\mathbf{z}, \mathbf{w}).$$

Therefore, regrouping the elements of each cluster leads to a loss of the information and a natural objective will be to search for the partitions that minimize this loss. This leads to the maximization of the criterion

$$F(\mathbf{z}, \mathbf{w}) = \phi^2(\mathbf{z}, \mathbf{w}). \qquad [I.5]$$

A similar development can be made starting from the mutual information

$$\mathcal{I}(I, J) = \sum_{i,j} p_{ij} \ln \frac{p_{ij}}{p_{i.}p_{.j}}.$$

As we will see in Chapter 4, the two criteria $\phi^2(\mathbf{z}, \mathbf{w})$ and $\mathcal{I}(\mathbf{z}, \mathbf{w})$ are very close and give similar results in most situations.

I.4.1.4. *Algorithms*

The optimization of the previous criteria is an NP-hard problem and heuristic algorithms must be used. Different approaches have been proposed. Alternated optimization algorithms are the most frequent approach: for instance, for criteria [I.2], [I.4] and [I.5], Govaert [GOV 77, GOV 83, GOV 95] developed three algorithms CROEUC, CROBIN and CROKI2, which alternate between row and column partitioning until the criterion reaches a local optimum. Bock [BOC 79] and Dhillon *et al.* [DHI 03] proposed similar algorithms, respectively, for continuous data and contingency table. These methods are fast and can process large data sets. Far less computation is required than for processing the two sets separately because, as we will see, the clustering is performed on reduced intermediate matrices of sizes $(n \times m)$ and $(d \times g)$.

Many other algorithms have been proposed: for instance, the sequential algorithm [POD 91], the genetic algorithm [HAN 00, NIE 05], simulated annealing [BRY 05] and tabu search [VON 09]. In [PUO 08] and [TIB 99], the authors compared these approaches with the use of classical clustering algorithms applied separately on the two sets. Furthermore, in some situations such as in bioinformatics, it is more interesting to divide the data matrix into blocks, which specify a unique partition for objects and a unique partition for variables. Specifically, the data matrix is divided into blocks where, conditionally to each class of objects, a different partition of the variables is allowed. Rocci and Vichi [ROC 08] proposed a generalized double $k$-means.

### I.4.2. *Two-mode hierarchical clustering*

The two-mode partitioning approach that we have seen can be extended to hierarchical clustering and, as for partitions, the process can be done separately or simultaneously.

#### I.4.2.1. *Separate clustering*

In a study on the joint use of classification and analysis of correspondence, Jambu [JAM 76] established research links between the two hierarchical classifications obtained from the two sets $I$ and $J$. The author used time-budgets where $I$ represents a set of population types while $J$ is a set of activity types undertaken by the population $I$. The simultaneous analysis is made on the two hierarchical models using the notion of contribution. In particular, the contributions of elements of $I$ classes built on $J$ and each element of $J$ classes built on $I$ are defined. Greenacre [GRE 88b] proposed the same approach and provided a simple graphical procedure that is useful in interpreting a significant chi-squared statistic of a contingency table. In a similar way, Camiz and Denimal [CAM 98] analyzed a two-way contingency table using two hierarchies obtained by a classical approach (Ward's method) on each set $I$ and $J$.

### I.4.2.2. *Simultaneous clustering*

Corsten and Denis [COR 90] proposed a two-mode hierarchical clustering. Toledano and Brousse [TOL 77] posed a similar problem: to simultaneously build homogeneous groups of individuals and groups of variables. Their objective is the search for two hierarchies checking this property. For this, they proposed an algorithm, called the double aggregation that seeks the best couple of lines or columns to be incorporated at each iteration. Eckes and Orlik [ECK 93] proposed an agglomerative algorithm for constructing a two-mode hierarchical classification. At each step of the process, the proposed algorithm merges biclusters whose fusion results in the smallest possible increase in an internal heterogeneity measure. This one is based on the variance within the respective cluster and the squared deviation of its mean from the maximum entry in the original data.

For binary data, a two-mode hierarchical clustering algorithm based on dissimilarity measures, derived from the latent block model and illustrated in Figure I.5, was developed by Jollois and Nadif [JOL 04]. Besides, in [JOL 03], a hybrid method jointly using two-mode partitioning and two-mode hierarchical clustering was proposed, enabling us to process large data sets. Both techniques derive from the latent block model described in details in Chapter 1.

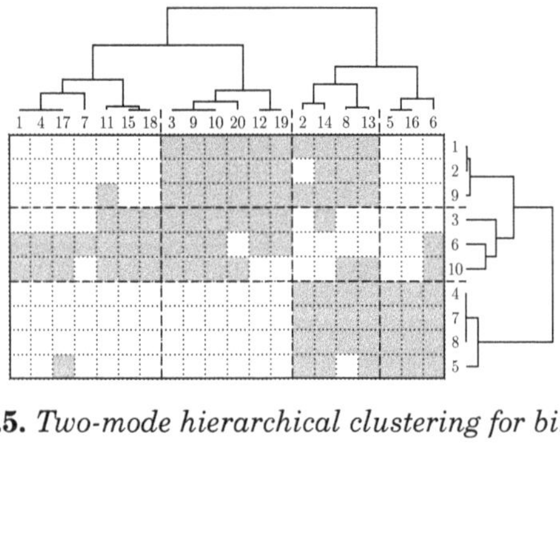


**Figure I.5.** *Two-mode hierarchical clustering for binary data*

### I.4.3. *Direct or block clustering*

In the earliest and most cited biclustering formulation, known as direct or block clustering, Hartigan [HAR 72] defined three families of clusters: the cluster of *responses* or observed values, i.e. a bicluster on the set of $I \times J$, the marginal cluster of objects on $I$ and the marginal cluster of variables (columns) on $J$. Thus, he suggested the following structures:

– The three-tree structure which requires that all three families of clusters on $I \times J$, $I$ and $J$ are trees.

– The partitioned responses structure which assumes that the clusters of response partition $I \times J$, but that the marginal row clusters and the marginal column clusters form trees on $I$ and $J$, respectively.

– The three partitions structure which requires that the three families are partitions of $I \times J$, $I$ and $J$ and which is the previous two-mode partitioning.

Note that the observed values must be comparable among variables as well as in the original data or after some previous normalization as discussed in section I.1.3. Then using a stepwise divisive method, Hartigan [HAR 72] developed an algorithm minimizing criterion [I.1]. Furthermore, Tibshirani *et al.* [TIB 99] added a backward pruning and devised a permutation-based method for deciding on the optimal number of blocks. Duffy and Quiroz [DUF 91] proposed another permutation-based algorithm for the same type of structures such that this approach can be extended to a wide variety of data, including matrices of categorical data. Besides, Eckes and Orlik [ECK 93] developed a hierarchical agglomerative algorithm to obtain a hierarchy of blocks.

### I.4.4. *Biclustering*

Since the late 1990s, biclustering has been the term most widely used in bioinformatics to identify co-clustering. In this situation, the structure is no longer limited to three families of clusters as suggested by Hartigan but generalizes the previous co-clustering structure. Its objective is to find a set of $K$ biclusters denoted as $B = \{B_1, \ldots, B_K\}$, where each bicluster $B_k$ is defined as follows:

$$B_k = (I_k, J_k) \text{ with } I_k \subseteq I \text{ and } J_k \subseteq J.$$

Different structures of biclusters exist [MAD 04] and some of them are illustrated in Figure I.6. For instance, S1 corresponds to four biclusters of different sizes with no overlap, S2 corresponds to four biclusters of the same size with overlap on two dimensions, S3 corresponds to four biclusters of the same size with a total overlap on one dimension and S4 corresponds to four biclusters of different sizes with multiple overlaps on one dimension.

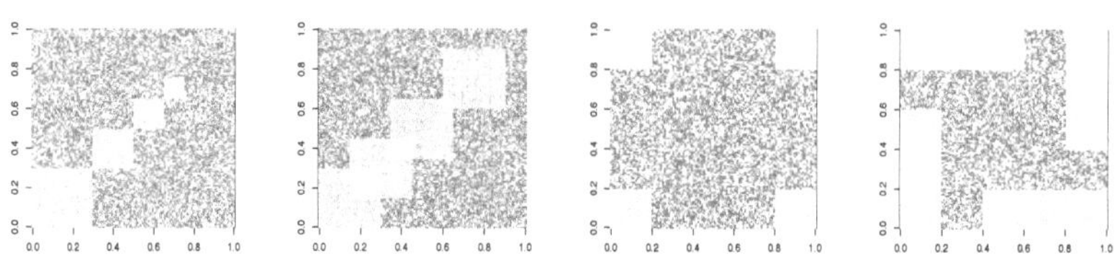


**Figure I.6.** *Examples of bicluster structures in artificial data sets*

In this context, consider the data matrix x where $I$ is the set of $n$ genes represented by $d$-dimensional vectors, $J$ is the set of $d$ conditions and $x_{ij}$ represents the expression level of gene $i$ under condition $j$. A bicluster $B_k$ is a submatrix of x defined by a subset of genes $I_k$ and a subset of conditions $J_k$. The biclustering aims, for instance, to identify groups of genes that show similar activity patterns under a specific subset of the experimental conditions. Note that a submatrix is considered as a bicluster if it presents a particular pattern. There is no definition of what these patterns are.

The choice of considering a submatrix as a bicluster is subjective and depends on the context. However, there are some basic patterns that can be used to identify a bicluster. They are called constant, additive and multiplicative models. In constant models, all values in a bicluster are equal. In additive and multiplicative models, there is an additive factor and a multiplicative factor between rows and columns, respectively. Biclusters can also be identified as a combination of these three models. Generally, the searched patterns are based on the correlation among the features or on the coherence among the values in the bicluster as discussed in Madeira's survey [MAD 04]. Biclusters can overlap on the rows and/or columns, or present a tree or checkerboard structure. This diversity in the nature of biclusters accounts for the fact that no biclustering algorithm can identify all types of biclusters.

Several biclustering algorithms have been proposed in the literature and applied to various domains. One of the most active domains is bioinformatics, especially microarray data analysis. The microarrays simultaneously measure the expression of a whole genome under different experimental conditions. Based on these microarray data, we can discover groups of genes with similar expression profiles over a subset of experimental conditions [GOR 02]. The hypothesis is that genes with similar expression profiles should have similar biological functions. The biclustering is also applied to drug activity data in order to associate common properties of chemical compound with common groups of features [LIU 03]. Two of the most popular biclustering methods used in bioinformatics are Chench and Church's algorithm and the plaid model.

Cheng and Church [CHE 00] were the first to propose a biclustering algorithm for microarray data analysis. Their algorithmic framework represents the biclustering problem as an optimization problem, defining a score for each bicluster candidate and developing heuristics to solve the constrained optimization problem defined by this score function. They consider that biclusters follow an additive model and use a greedy iterative search to minimize the mean square residue (MSR). This algorithm identifies biclusters $B_k = (I_k, J_k)$ one by one by using the score

$$MSR(B_k) = \frac{1}{|I_k| \times |J_k|} \sum_{i,j|i \in I_k, j \in J_k} (x_{ij} - \bar{x}_{i.} - \bar{x}_{.j} + \bar{x}_{..})^2,$$

where $\bar{x}_{i.} = \frac{1}{|I_k|} \sum_{i|i \in I_k} x_{ij}$, $x_{.j} = \frac{1}{|J_k|} \sum_{j|j \in J_k} x_{ij}$ and $\bar{x}_{..} = \frac{1}{|I_k| \times |J_k|} \sum_{i,j|i \in I_k, j \in J_k} x_{ij}$.

Lazzeroni and Owen [LAZ 00] proposed the popular plaid model that was improved by Turner *et al.* [TUR 05]. They assume that biclusters are organized in layers and follow a given statistical model incorporating additive two-way ANOVA models. The values of $x_{ij}$ belonging to a bicluster $B_k$ (layer) depend on four parameters: a constant $\mu_0$ describing the background layer, $\mu_k$ the average of $B_k$, $\alpha_{ik}$ and $\beta_{jk}$ allowing us to identify, respectively, a subset of genes and examples with identical responses. The values of $x_{ij}$ are therefore represented as

$$x_{ij} = \mu_0 + \sum_{k=1}^{K} \sum_{i,j|i \in I_k, j \in I_k} (\mu_k + \alpha_{ik} + \beta_{jk}).$$

The search approach is iterative: once $K - 1$ layers (biclusters) have been identified, the $K$-th bicluster that minimizes a merit function depending on all layers is selected.

Tanay *et al.* [TAN 02] developed statistical-algorithmic method for bicluster analysis (SAMBA), an approach based on the graph theory coupled with statistical modeling of the data. Prelic *et al.* [PRE 06] made a comparative study of different biclustering methods for gene expression. They used the Bimax algorithm, a very simple divide and conquer approach as a reference to investigate the usefulness of different biclustering algorithms. They concluded that Bimax produces results similar to those of more complex methods. Note that in such situations, no criterion can be defined and this approach has led to many heuristics: [OYA 01] (ping-pong algorithm), [IHM 02, IHM 04a, IHM 04b, BEN 03, BER 03, TAN 05, TCH 06]. Showing a connection between spectral partition and crossing minimization, Ahmed and Whokhar [AHM 07] developed an efficient biclustering.

Gupta and Aggarwal [GUP 10] introduced the use of mutual information to find relevant biclusters in gene expression data. Mitra and Banka [MIT 06] pointed out that biclustering in microarray data consists of finding sub-matrices, i.e. maximal subgroups of genes and subgroups of conditions where the genes exhibit highly correlated activities over a range of conditions. They have shown that these two objectives are mutually conflicting and they proposed a multiobjective evolutionary biclustering framework by incorporating local search strategies. Cho and Dhillon [CHO 08] proposed a fast biclustering algorithm that minimizes the sum squared residue and generates a checkerboard structure in the data matrix. Erten and Sözdinler [ERT 10] developed their localization procedure that improves the performance of a greedy iterative biclustering algorithm and applied it to microarray data sets. Several other methods were proposed in the literature, a good survey of biclustering methods for biological data analysis was published by Madeira and Oliveira [MAD 04], they enumerated more than 15 algorithms used in this context. Furthermore, to improve the performance of these biclustering methods, Hanczar and Nadif [HAN 11, HAN 12b] adapted the bagging approach to biclustering problems. The principle consists of generating a set of biclusters and aggregating the results. On simulated and real data sets, this approach appears effective.

### I.4.5. *Other structures and other aims*

#### I.4.5.1. *Block diagonal structure*

In the two-mode partitioning algorithms, constraints can be added to obtain a structure of diagonal blocks after row and column reordering: the row partition and the column partition must have the same number of clusters ($g = m$) and the diagonal biclusters $(I_k, J_k)$ must take a different form from the other biclusters. For instance, in the binary data case, the diagonal biclusters will be composed primarily of value 1 and other biclusters of 0. Criterion [I.4] with the constraint that the matrix a $= (a_{k\ell})$ is the identity matrix is well adapted to this situation and was used, for instance, by Garcia and Proth [GAR 86] in a group technology application.

This type of approach is also known as *block seriation*. The techniques of seriation, applied in various fields such as sociology, archaeology, botany and zoology, amount to finding a permutation of rows associated with a permutation of columns allowing us to extract the data from a latent order. Block seriation corresponds to a particular approach to this problem and in this context, Marcotorchino [MAR 91a] proposed a solution using linear programming. The BEA [MCC 72, ARA 90] can also be used to obtain a block diagonal structure. Recently, a normalized generalized modularity criterion for binary and categorical data with the aim of co-clustering, was proposed by Labiod and Nadif [LAB 11b]. For its maximization, the authors developed a spectral algorithm and studied the problem of the number of blocks.

Various other approaches have been proposed: Pensa *et al.*, [PEN 05] developed an algorithm leading to a block diagonal structure with the availability of overlapping. In [DHI 01], modeling the document collection as a bipartite graph between documents and words, the author posed the simultaneous clustering problem as a bipartite graph partitioning problem and obtained the clusters by using the second left and right singular vectors of the singular value decomposition of an appropriately scaled word-document matrix.

#### I.4.5.2. *Different column clustering for each row cluster*

Some authors [POL 02, ROC 08] suggested treating the two-mode clustering by first clustering the rows and then for each row cluster, clustering the columns. For instance, in the partitioning situation, conditionally to each class of row partition, a different partition of columns is allowed.

#### I.4.5.3. *Multi-way data*

More complex situations exist where the data take the form of a multidimensional array instead of a matrix. For example, multiple variables measured on a set of objects over time, or contingency tables defined on more than two categorical data are examples of array-valued or multi-way data. Some of the previous approaches have been extended to this situation. For instance, Ambroise and Govaert [AMB 02] proposed a clustering of this multiway data along all its dimensions simultaneously using a model-based strategy, and Peng *et al.* [PEN 08] proposed the subspace clustering algorithm.

#### I.4.5.4. *Principal component analysis and correspondence analysis*

The principal component analysis and its variants such as correspondence analysis methods are also applicable to matrices defined on two sets simultaneously and obtain results on these two sets. It is also possible to show that certain methods of cross-classification can be seen as principal component analysis under constraints.

## I.5. Model-based co-clustering

Clustering methods can be roughly divided into two categories. The first is based on a choice of some distance or distortion measure among the data points, which presumably reflects some background knowledge of the data. For most problems, a proper choice of the distance measure can be the main practical difficulty through which much of the arbitrariness of the results can enter. Another class of methods is based on statistical assumptions relating to the origin of the data. Such assumptions enable the design of a statistical model where the model parameters are then estimated based on the given data and, in this situation, basing cluster analysis on mixture models has become a classical and powerful approach. The works of Banfield and Raftery [BAN 93], Celeux and Govaert [CEL 92, CEL 93] and McLachlan [MCL 82] are recent examples, among many others, of this point of view. For co-clustering, even though it is less common and more recent, various models have been proposed.

For instance, Rooth [ROO 95] proposed a probabilistic model for block clustering of contingency tables with a block diagonal structure (see section I.4.5.1 for a description of this structure) and proposed an algorithm that uses formulas similar to the Baum–Welch re-estimation formulas for hidden Markov models. Hartigan [HAR 00] used probabilistic models to perform block clustering on binary data. Nowicki and Snijders [NOW 01] proposed a stochastic block structures model that builds a mixture model for stochastic relationships among objects and identifies the latent cluster via posterior inference. Kemp *et al.* [KEM 06] proposed an infinite relational model that discovers stochastic structure in relational data in the form of binary observations. Airoldi *et al.* [AIR 08] proposed a mixed membership stochastic block model that relaxes the single-latent-role restriction in stochastic block structure models.

Govaert and Nadif [GOV 03, GOV 08] proposed the latent block model for binary data (binary latent block model) and they extended this model to contingency table (Poisson latent block model) in [GOV 10]. In each case, variational approach, also called a mean-field approximation, of expectation-maximization (EM) algorithm [DEM 77] was used to estimate the parameters and the partitions. Lashkari and Golland [LAS 09] proposed the same generative model and also used a variational approach. They showed that this model has modeling assumptions in common with the Bregman co-clustering of Banerjee *et al.* [BAN 07]. In analyzing continuous data in gene expression context, Jagalur *et al.* [JAG 07] used the latent block model where the conditional distributions, knowing the row and the column clusters, are Gaussian. They applied the variational EM and the classification EM (CEM) algorithms. They also proposed a sequential optimization algorithm of the criterion defined in the CEM approach of the latent block model. For text categorization, Takamura and Matsumoto [TAK 02] proposed a greedy algorithm to estimate the parameters and the two partitions of the latent block model simultaneously (classification approach) and used the Akaike criterion as a stopping rule. In the contingency table situation, Hofmann and Puzicha [HOF 99a] presented different clustering models and in the two-sided clustering situation, the modeling assumptions are equivalent to the relations obtained by the Poisson latent block model. They proposed an approximate EM algorithm using the variational approach and in the classification approach, they used a mutual information criterion.

To predict customer product preference in a market application, Deodhar and Ghosh [DEO 07] proposed a model-based co-clustering model that can be viewed as an extension of the latent block model, taking into account customer and product attributes. The proposed algorithm interleaves the clustering of customers and products and the construction of prediction models.

Different Bayesian approaches to this kind of model have recently been proposed. Shan and Banerjee [SHA 08] and Dijk *et al.* [DIJ 09] developed a Bayesian approach of the latent block model to estimate the parameters. The first proposed a variational approach while the latter used Gibbs sampling algorithms. Similar works on the latest block model have been developed by Meeds and Roweis [MEE 07], which have showed how these models can easily take into account missing data and are robust to high rates of missing data. Starting from a probabilistic model on a contingency table, in [POI 08], the authors also used a Bayesian approach to define a clustering criterion and proposed a greedy two-mode clustering algorithm to optimize this criterion.

In the collaborative filtering context, Kleinberg and Sandler [KLE 08] proposed a mixture model, and a statistical model of collaborative filtering similar to the Bernoulli latent block model proposed by Ungar and Foster [UNG 98]. For estimating the model parameters, they have developed different methods including variations of the $k$-means algorithm and Gibbs sampling. Shafiei and Millios [SHA 06] extended these approaches and proposed a model-based overlapping co-clustering able to work with any regular exponential family distribution.

Model-based approaches have also been used to treat the situation described in section I.4.5.2, where different column clusterings are used for each row cluster. In the analysis of gene expression data, Pollard and van der Laan [POL 02] proposed a probabilistic model and used the non-parametric bootstrap method to assess the variability of the estimator. In document and word clustering, Li and Zha [LI 06] used mixtures of Poisson distribution to model the multivariate distribution of the word counts in the document within each class.

## I.6. Outline

In this book, we mainly deal with the two-mode partitioning under different approaches but with particular attention to a probabilistic approach. This book is organized as follows.

– Chapter 1 concerns clustering in general and model-based clustering in particular. We briefly review classical clustering methods and focus on the mixture model. We present and discuss the use of different mixtures adapted to different types of data. The algorithms used are described, and related works with different classical methods are presented and discussed. This chapter will be useful for tackling the problem of co-clustering under the mixture approach.

– Chapter 2 is devoted to the latent block model proposed in the mixture approach context. We discuss this model in detail and show its significance in co-clustering. Various algorithms are presented in a general context.

– Chapter 3 focuses on binary and categorical data. It presents the appropriated latent block mixture models in detail. Variants of these models and of algorithms are presented and illustrated by examples.

– Chapter 4 is devoted to the contingency table. Mutual information, phi-squared coefficient and model-based co-clustering are studied. Models, algorithms and connections among different approaches are described and illustrated.

– Chapter 5 presents the case of continuous data. In the same manner, the different approaches used in the previous chapters are extended to this case.

# Chapter 1

# Cluster Analysis

## 1.1. Introduction

*Cluster analysis* or *clustering*, which is an important tool in a variety of scientific areas including pattern recognition, information retrieval, microarrays and data mining, is a family of exploratory data analysis methods that can be used to discover structures in data. These methods seek to obtain a reduced representation of the initial data and, along with principal component analysis, factor analysis and multidimensional scaling, are one form of data reduction. The aim of cluster analysis is the organization of the set into *homogeneous classes* or *natural classes*, in a way which ensures that objects within a class are similar to one another. For example, in statistics, cluster analysis can identify several populations within a heterogeneous initial population, thereby facilitating a subsequent statistical study; in natural science, the clustering of animal and plant species, first proposed by Linnaeus (an 18th-Century Swedish naturalist), is a famous example of cluster analysis; in the study of social networks, clustering may be used to recognize communities within large groups of people and, at a more general level, simply naming of objects can be seen as a form of clustering.

The attempt to formally define clustering, as a basis for an automated process, raises a number of questions. How can we define the objects (elements, cases, individuals or observations) to be classified? What is a cluster? How are clusters structured? How can different partitions be compared? Most often, the first step consists of defining the notion of proximity, a measure of closeness that can be similarity, dissimilarity or distance, among the objects to be clustered: two objects are close when their dissimilarity or distance is small or their similarity is large. Sometimes these proximities are the form in which the data naturally occur. In most clustering problems, however, each of the objects under investigation will be described by a set of *variables* or *attributes*, and the first step, possibly the most important, in clustering is to define these proximities. Then, a numerical function, usually known as a *criterion*, measuring the homogeneity of the clusters must be defined.

A classical example of a criterion used when the objects $\boldsymbol{x}_1, \ldots, \boldsymbol{x}_n$ are described by $d$ continuous variables is the *within-group sum of squares*, also called *within-group inertia*. In this situation, each individual being characterized by a vector $\boldsymbol{x}_i = (x_{i1}, \ldots, x_{id})$, the data take the form of a matrix x of dimension $(n, d)$ defined by values $x_{ij}$, where $i$ belongs to a set $I$ of $n$ observations and $j$ belongs to a set $J$ of $d$ continuous variables. The within-group sum of squares can therefore be written as

$$I_W(\mathbf{z}) = \frac{1}{n}\sum_{i,k} z_{ik}\mathrm{d}^2(\boldsymbol{x}_i, \overline{\boldsymbol{x}}_k) = \frac{1}{n}\sum_{i,k} z_{ik}||\boldsymbol{x}_i - \overline{\boldsymbol{x}}_k||^2, \qquad [1.1]$$

where $\overline{\boldsymbol{x}}_k$ is the mean vector of the $k$th cluster and d is the Euclidean distance. Using the within-group covariance matrix

$\mathbf{S}_W = \frac{1}{n}\sum_k z_{.k}\mathbf{S}_k$, where $z_{.k}$ is the size of the $k$th cluster and $\mathbf{S}_k$ is the covariance matrix of the $k$th cluster

$$\mathbf{S}_k = \frac{1}{z_{.k}}\sum_i z_{ik}(\boldsymbol{x}_i - \overline{\boldsymbol{x}}_k)(\boldsymbol{x}_i - \overline{\boldsymbol{x}}_k)^t,$$

this criterion can also be written as $I_W(\mathbf{z}) = \text{trace}(\mathbf{S}_W)$. The closer the within-group sum-of-squares criterion is to 0, the more homogeneous the partition will be. In particular, this criterion will be equal to 0 for a partition where each object is a cluster.

The problem can then appear very simple: from the finite set of partitions, select the partition that optimizes the numerical criterion. Unfortunately, the number of partitions is too large for them to be enumerated in a realistic time frame, because of combinatorial complexity. Generally, heuristics are used that, rather than giving the best solution, give a "good" solution close to the optimal solution and lead to local optimization. For instance, the two most commonly used clustering algorithms, namely the $k$-means algorithm for obtaining partitions and Ward's hierarchical clustering method for obtaining hierarchies, use the within-group sum of squares criterion $\text{trace}(\mathbf{S}_W)$ derived from the within-group covariance matrix $\mathbf{S}_W$, and used the Euclidean metric as a measure of proximity.

In recent years, what used to be an algorithmic, heuristic and geometric focus has tended to give way to a more statistical approach using probabilistic clustering models to formalize the intuitive notion of a natural class [BOC 89]. This approach allows precise analysis and can provide a statistical interpretation of certain metrical criteria whose different variants are not always clear (such as the within-group sum of squares criterion $\text{trace}(\mathbf{S}_W)$), as well as yielding new variants corresponding to precise hypotheses. It also represents a formal framework for tackling difficult problems such as determining the number of classes or validating the obtained clustering structure. We should bear in mind that in many cases the set to be segmented is merely a sample drawn from a much larger population, and that the conclusions drawn from clustering the sample are to be extrapolated to the entire population. Here, clustering becomes meaningless in the absence of a probabilistic model justifying this extrapolation. All probabilistic approaches to clustering first assume that the data represent a random sample $x_1, \ldots, x_n$ from among a population, and then use an analysis of the probability distribution of this population to define a clustering. A number of different probabilistic clustering methods have been proposed, but the most traditional approach is the use of mixture models, which forms the main subject of this chapter.

Section 1.2 presents a brief review of the main approaches to clustering. Sections 1.3 and 1.4 deal with, respectively, the probability mixture models and the EM algorithm, the standard tool for estimating the parameters of such models. Section 1.5 describes how clustering may be carried out using a mixture model. The four subsequent sections describe several classical situations including Gaussian mixture models for continuous variables, and the latent class model for binary variables, categorical variables and contingency table, and in section 1.10, we study the implementation of these different methods.

## 1.2. Miscellaneous clustering methods

### 1.2.1. *Hierarchical approach*

In this section, it will generally be assumed that all the relevant relationships within the set to be classified are summarized by a dissimilarity d. The aim of hierarchical methods is to construct a sequence of partitions of a set varying from partitions of singletons to the whole set. There are two principal approaches. The divisive approach starts with just one cluster containing all the objects. In each successive iteration, clusters are split into two or more further clusters, usually until every object is alone within a cluster. Note that other stop conditions can be used, and the division into clusters is governed by whether or not a particular property is satisfied. For example, in taxonomy, animals may be separated into vertebrates and invertebrates. The agglomerative approach, in contrast to the divisive approach, starts out from a set of $n$ clusters, with each object forming a singleton cluster. Then, in each successive iteration, the closest clusters are merged until just one cluster remains. Using a dissimilarity $D$ among groups, the closest clusters are the two clusters that are the *closest* with respect to $D$. According to the definition of $D$, several agglomerative criteria exist, but the most commonly used are the single linkage or nearest-neighbor criterion [SIB 73], the complete linkage or furthest-neighbor criterion [SOR 48] and the average linkage criterion [SOK 58]. When the objects are described by continuous variables, it is also possible to use Ward's method.

### 1.2.2. *The $k$-means algorithm*

This section deals with the $k$-means algorithm, which is the classical method in partitional clustering when the data are a set of objects $x_1, \ldots, x_n$ described by $d$ continuous variables. The objective of partitional (or non-hierarchical clustering) is to define the partition of a set of objects into clusters, that the objects in a cluster are more "similar" to each other than to objects in other clusters. Starting from $g$ initial cluster centers, the $k$-means algorithm involves the two following steps up to the convergence: assign each object in $\Omega$ to the nearest cluster center; use the centroids of the different clusters as the new cluster centers. It can easily be shown that this algorithm yields a stationary sequence of partition decreasing the within-group sum-of-squares criterion.

The term $k$-means actually covers a whole family of methods, and the algorithm previously described is only one example. Bock [BOC 07] has carried out an interesting survey of some historical issues related to the $k$-means algorithm. We can cite Dalenius [DAL 50, DAL 51], Lloyd's algorithm [LLO 57] in the context of scalar quantization in the one-dimensional case and Steinhaus [STE 56] for data in $\mathbb{R}^d$ in the multidimensional case. Different strategies have been used: first, we have batch algorithms that process all the objects of the sample at each iteration, and incremental algorithms that process only one object at each iteration. Second, we have on-line or off-line training. In on-line training, each object is discarded after it has been processed (on-line training is always incremental). In off-line training, all the data are stored and can be accessed repeatedly (batch algorithms are always off-line). The term *sequential* is ambiguous, referring sometimes to incremental algorithms and sometimes to on-line learning. On-line $k$-means variants are particularly suitable when not all the data to be classified are available at the outset. The $k$-means algorithm previously described is an example of a batch algorithm. This batch version, which is the one used most often, was proposed by Forgy [FOR 65], Jancey [JAN 66] or Linde *et al.* [LIN 80] in vector quantization. The algorithm of MacQueen [MAC 67], who was the first to use the name "$k$-means", and the *neural gas* algorithm [MAR 91b] are examples of on-line $k$-means.

If the aim is to find the partition minimizing the within-group sum-of-squares criterion, the $k$-means algorithm does not necessarily provide the best result, but simply a sequence of partitions, the value of whose criterion will decrease, thus giving a local optimum. Since, in practice, convergence is reached very quickly (often in fewer than 10 iterations even with a large data set), the user can run $k$-means several times, with different random initializations and retain the best partition, i.e. that optimizes the criterion.

If the number of clusters is not known, several solutions are possible to solve this very difficult problem. For example, the best partition is sought for several numbers of classes and the number of classes are selected by choosing an elbow on the scree plot. It is also possible to add additional constraints relating, for example, to the number of objects by cluster or to the volume of a cluster (see, for instance, the Isodata algorithm [BAL 67]). Finally, there are other approaches using statistical methods, such as hypothesis tests or selection model criteria. This last approach, apparently the most interesting, consists of penalizing the criterion by a function depending on the number of classes, making the criterion "independent" of this number of classes. For instance, minimizing the within-group sum of squares can be viewed as maximizing a likelihood (see section 1.6), and selection model criteria such as the Akaike information criterion (AIC) or the Bayesian information criterion (BIC) can be used to determine the number of clusters.

Finally, it can be interesting to use $k$-means and Ward's method simultaneously: the two methods are similar in that they both attempt to minimize the within-group sum-of-squares criterion. This leads us to propose strategies using the two approaches, such as, for example, the hybrid method proposed by Wong [WON 82].

### 1.2.3. *Other approaches*

The *dynamic cluster method* proposed by Diday [DID 71, DID 76] is a generalization of the $k$-means algorithm based on the quite powerful idea that the cluster centers are not necessarily centroids of clusters in $\mathbb{R}^d$ and to replace them by centers that may take a variety of forms, depending on the problem to be solved. The $k$-medoids algorithm ([KAU 87]) is a typical dynamic cluster method where the cluster centers are objects of the set to cluster. Different versions have been proposed: partition around medoids (PAM) ([KAU 90]) and CLARANS ([NG 94]), which are more efficient for large volumes of data. This method is particularly well adapted when the data are given as a dissimilarity matrix $d$. The $k$-modes algorithm [HUA 97, NAD 93] for categorical data is another example in which the centers are vectors of categories. A final example (*adaptive distance*, [DID 74, DID 77]), in which each center is a pair of point and distance, determines partition and distance simultaneously allowing the shapes of clusters to be taken into account.

Fuzzy clustering [RUS 69], developed to handle the notion of overlapped clusters, generalizes the classical approach in clustering by assuming that each element $x_i$ can belong to more than one cluster with different levels. A fuzzy partition can therefore be represented by a fuzzy classification matrix $\mathrm{c} = \{c_{ik}\}$ satisfying the following conditions: $\forall i, k,\ c_{ik} \in [0, 1]$, $\forall k,\ \sum_i c_{ik} > 0$ and $\forall i,\ \sum_k c_{ik} = 1$. Bezdeck [BEZ 81] proposed the *fuzzy $k$-means* algorithm, which can be viewed as a fuzzy version of $k$-means. The parameter estimation of a mixture model (see section 1.3) can also be viewed as a fuzzy clustering, and the associated EM algorithm is a more statistically formalized method that includes the notion of partial membership in classes. It has better convergence properties and is in general preferred to fuzzy $k$-means.

The *self-organizing map* (SOM) or *Kohonen map* [KOH 82] was first inspired by the adaptive formation of topology-conserving neural projection in the brain. Its aim is to generate a mapping of a set of high-dimensional input signals onto a one- or two-dimensional array of formal neurons. Each neuron becomes representative of some input signals, such that the topological relationship among input signals in the input space is reflected, as faithfully as possible, in the arrangement of the corresponding neurons in the array (also called output space). When using this method for clustering, it is possible either to match each neuron with a unique cluster or to match many neurons to one cluster. In the latter case, the Kohonen algorithm produces a reduced representation of the original data set, and clustering algorithms may operate on this new representation. In the SOM literature, we refer to the clusters by the nodes or neurons, each of which has a weight in $\mathbb{R}^d$. The weights refer to the cluster means. The principal advantage of SOM is that it preserves the topology clustering. Generally, the neurons are arranged as a one- or two-dimensional rectangular grid preserving relations among the objects, also referred to as units. SOM is therefore a useful tool for visualizing clusters and evaluating their proximity in a reduced space. The Kohonen map can be viewed as an extension of the on-line $k$-means algorithm and, like $k$-means, requires the number of clusters (nodes of the grid) to be fixed and initial values to be selected. Different strategies can be used but initialization using principal component analysis (PCA) would appear to be an attractive and interesting approach.

High-dimensional data present a particular challenge to clustering algorithms. This is because of the so-called curse of dimensionality that leads to the sparsity of the data: in high-dimensional space, all pairs of points tend to be almost equidistant from one another. As a result, it is often unrealistic to define distance-based clusters in a meaningful way. Usually, clusters cannot be found in the original feature space because several features may be irrelevant for clustering, owing to correlation or redundancy. However, clusters are usually embedded in lower dimensional subspaces, and different sets of features may be relevant for different sets of objects. Thus, objects can often be clustered differently in subspaces varying from the original feature space. Different approaches have been proposed. *Subspace clustering* seeks to find clusters in different subspaces within a data set. An example of this kind of approach is the CLIQUE algorithm [AGR 98]. It is a density-based method that can automatically find subspaces of the highest dimensionality such that high-density clusters exist in those subspaces. *Subspace ranking* aims at identifying all subspaces of a (high-dimensional) feature space that contain interesting clustering structures. The subspaces should be ranked according to this interestingness. *Projected clustering* is a method whereby the subsets of dimensions selected are specific to the clusters themselves. We can also cite the high-dimensional data clustering (HDDC) approach of Bouveyron [BOU 07]. In this approach, a family of Gaussian mixture models designed for high-dimensional data combining the ideas of subspace clustering and parsimonious modeling are used to develop a clustering method based on the EM algorithm.

*Kernel clustering* methods have attracted much attention in recent years [FIL 08]. They involve transforming a low-dimensional input space into a high-dimensional kernel-deduced feature space in which patterns are more likely to be linearly separable [SCH 02]. They have certain advantages when handling nonlinear separable data sets. We can also cite *spectral clustering* techniques that make use of the spectrum of the similarity matrix of the data to perform dimensionality reduction for clustering in fewer dimensions, and more generally, graph clustering techniques.

In this section, different approaches to clustering, essentially classical and generally based on numerical criteria, have been reviewed. Unfortunately, defining these criteria and using them successfully is not always easy. To overcome these difficulties exist other approaches, such as the mixture model approach, which is undoubtedly a very useful contribution to clustering. It offers considerable flexibility, gives a meaning to certain criteria and sometimes leads to replacing criteria with new criteria with fewer drawbacks. In addition, it provides solutions to the problem of the number of clusters. The next section describes this approach.

## 1.3. Model-based clustering and the mixture model

The clustering methods described in the previous two sections are mainly heuristic techniques derived from empirical methods, usually optimizing measurement criteria. Implementing these solutions entails choosing not only a metric reflecting the dissimilarity among the objects in the set to be segmented, but also a criterion deriving from this metric capable of measuring the degree of cohesion and separation among classes. A rapid perusal of the lists of metrics and criteria proposed in the clustering literature will be enough to convince most readers that these are not easy choices. A number of different probabilistic clustering methods have been proposed, but the use of finite mixture densities, which provides a sensible statistical model for the clustering process, is now widespread.

Since their first use by Newcomb in 1886 for the detection of outlier points, and then by Pearson in 1894 to identify two separate populations of crabs, finite mixtures of distributions have been employed to model a wide variety of random phenomena. These models assume that measurements are taken from a set of individuals, each of which belongs to one of a number of different classes, while any individual's particular class is unknown. We might, for instance, know the sizes of fish in a sample, but not their sex, which is difficult to ascertain. Mixture models can thus address the heterogeneity of a population, and are especially well suited to the problem of clustering. This is an area where much research has been done. McLachlan and Peel's book [MCL 00] is a highly detailed reference for this domain that has seen considerable developments over the last few years. We will first briefly recall the model and the problems of estimating its parameters.

Finite mixture models, which assume that every class is characterized by a probability distribution, are highly flexible models that can take account of a variety of situations including heterogeneous populations and outlier elements. Because of the EM algorithm, which is particularly well suited to this kind of context, a number of mixture models have been developed in the field of statistics, and the use of mixture models in clustering has been studied by authors including Scott and Symons [SCO 71], Marriott [MAR 75], Symons [SYM 81], McLachlan [MCL 82] and McLachlan and Basford [MCL 88]. The mixture model approach is attractive for several reasons. It corresponds to our intuitive idea of a population composed of several classes, it is strongly linked to reference methods such as the $k$-means algorithm and it is able to handle a wide variety of special situations in a more or less natural way. It is this approach that forms the subject of this chapter.

In a finite probability mixture model, the data $\mathrm{x} = (\boldsymbol{x}_1, \ldots, \boldsymbol{x}_n)$ are taken to constitute a sample of $n$ independent instances of a random variable $\boldsymbol{X}$ in $\mathbb{R}^d$. The productivity density function (pdf) can be expressed as

$$f(\boldsymbol{x}_i) = \sum_k \pi_k f_k(\boldsymbol{x}_i), \quad \forall i \in I,$$

where $g$ is the number of components, $f_k$ are the pdf of each component and $\pi_k$ are the mixture proportions ($\pi_k \in ]0,1[ \ \forall k$ and $\sum_k \pi_k = 1$). The principle of a mixture model is to suppose, given the proportions $\pi_1, \ldots, \pi_g$ and the distributions $f_k$ of each class, that the data are generated according to the following mechanism:

– z: each individual is allotted to a class according to a categorical distribution with parameters $\pi_1, \ldots, \pi_g$;

– x: each $\boldsymbol{x}_i$ is assumed to arise from a random vector with pdf $f_k$.

In addition, it is usually assumed that the components' pdf $f_k$ belong to a parametric family of pdf $f(., \alpha)$. The pdf of the mixture can therefore be written as

$$f(\boldsymbol{x}_i, \boldsymbol{\theta}) = \sum_k \pi_k f(\boldsymbol{x}_i; \boldsymbol{\alpha}_k), \quad \forall i \in I,$$

where $\boldsymbol{\theta} = (\pi_1, \ldots, \pi_g, \boldsymbol{\alpha}_1, \ldots, \boldsymbol{\alpha}_g)$ is the parameter of the model. For example, the pdf of a mixture model for two univariate Gaussian distributions of variance 1 in $\mathbb{R}$ is written as

$$f(x_i; \pi, \mu_1, \mu_2) = \pi\varphi(x_i; \mu_1, 1) + (1 - \pi)\varphi(x_i; \mu_2, 1),$$

where $\varphi(.; \mu, \sigma^2)$ is the pdf of the univariate Gaussian distribution of mean $\mu$ and variance $\sigma^2$. Figure 1.1 uses the pdf obtained from a mixture of three Gaussian components in $\mathbb{R}^2$ to illustrate this concept of a probability mixture.

Much effort has been devoted to the estimation of parameters for the mixture model, following the work of Pearson, whose use of the method of moments to estimate the five parameters $(\mu_1, \mu_2, \sigma_1^2, \sigma_2^2, \pi)$ of a univariate Gaussian mixture model with two components required him to solve polynomial equations of degree nine. There have been a number of studies [MCL 88, TIT 85] and different estimation methods have been envisaged. Apart from the method of moments, we also find graphic methods, the maximum likelihood method and Bayesian approaches. In this chapter, we will restrict ourselves to examining the maximum likelihood method using the EM algorithm, which is currently the most widely used. Before examining this method in section 1.4, we first draw the reader's attention to certain difficulties which the estimation of parameters of a mixture model presents.

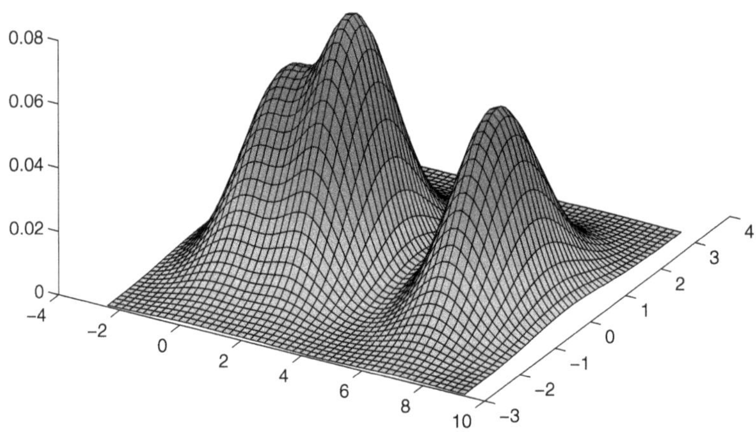


**Figure 1.1.** *Gaussian mixture in* $\mathbb{R}^2$

In certain situations, such as in the case of the crabs data previously described where the idea of the component has a precise physical basis, the number of components may be completely determined. Most often, however, the number of components is not known and must itself be estimated. It should be noted that if the number of components is taken to be an additional parameter, the mixture model may be seen as a semi-parametric compromise between a classical parametric estimation problem, when the number of components corresponds to a fixed constant, and a non-parametric estimation problem, in this case via the kernel method, when the number of components is equal to the size of the sample. We assume from here onwards that number $g$ of components is known, and later we will look at the proposed solutions for making this difficult choice.

If the problem is to be of any interest, the pdf of the mixture needs to be identifiable, which means that any two mixtures whose densities are the same must have the same parameters. A number of studies have addressed this problem and several difficulties arise. The first difficulty is due to the numbering of the classes. For example, in the case of a mixture with two components, the parameters $((\pi_1, \pi_2), (\alpha_1, \alpha_2)$ and $((\pi_2, \pi_1), (\alpha_2, \alpha_1)$, although different, obviously yield the same pdf: a mixture is consequently never identifiable. The difficulties to which this situation gives rise will depend on the estimation algorithms. In the case of the EM algorithm that we will use, it simply does not matter, however, this cannot be said of the Bayesian approach, where this situation is known as the "*switching problem*". The second, considerably more awkward, difficulty may arise from the very nature of the component pdf. It may easily be established that a mixture of uniform or binomial distributions is not identifiable. Mixtures of Gaussian, exponential and Poisson distributions, however, are identifiable.

## 1.4. EM algorithm

Maximizing the log-likelihood of a mixture model

$$\mathrm{L}(\boldsymbol{\theta}) = \log\left(\prod_i \sum_k \pi_k f(\boldsymbol{x}_i, \boldsymbol{\alpha}_k)\right)$$

leads to likelihood equations that usually have no analytical solution. It may, nevertheless, be shown that if the parameter

$\boldsymbol{\alpha}_k$ is a vector of real numbers $\alpha_{kr}$, the solution of these likelihood equations must satisfy

$$\pi_k = \frac{1}{n}\sum_i \widetilde{z}_{ik} \ \forall k \quad \text{and} \quad \sum_i \widetilde{z}_{ik}\frac{\partial \log f_k(\boldsymbol{x}_i, \boldsymbol{\alpha}_k)}{\partial \alpha_{kr}} = 0 \ \forall k, r \qquad [1.2]$$

$$\text{with} \quad \widetilde{z}_{ik} = \frac{\pi_k f_k(\boldsymbol{x}_i, \boldsymbol{\alpha}_k)}{\sum_{k'} \pi_{k'} f_{k'}(\boldsymbol{x}_i, \boldsymbol{\alpha}_{k'})}. \qquad [1.3]$$

These equations suggest the following iterative algorithm: (1) start from an initial solution $\theta$; (2) calculate the values $\widetilde{z}_{ik}$ from this parameter using equation [1.3]; (3) update the parameter $\theta$ on the basis of these values $\widetilde{z}_{ik}$ using equations [1.2]; continue from (2). If this algorithm converges, therefore, the fixed point obtained will satisfy the likelihood equations. The procedure corresponds, in fact, to the application of the Dempster *et al.* [DEM 77] EM algorithm to the mixture model. Before describing this algorithm, we will define the concept of complete data on which it relies.

### 1.4.1. *Complete data and complete-data likelihood*

At the outset, we consider that the observed data $\mathrm{x}$ correspond to what is merely a partial knowledge of unknown data $\mathrm{y}$ that are termed *complete data*, the two being linked by a function $\mathrm{x} = T(\mathrm{y})$. The complete data might, for instance, be of the form $\mathrm{y} = (\mathrm{x}, \mathrm{z})$, in which case $\mathrm{z}$ is known as *missing information*. This idea of complete data may either be meaningful for a model, which is the case for the mixture model, or it may be completely artificial. The likelihood $f(\mathrm{y}; \boldsymbol{\theta})$ calculated from these complete data is termed *complete-data likelihood* or, in the case of the mixture model, *classification likelihood*. Starting from the equation $f(\mathrm{y}; \boldsymbol{\theta}) = f(\mathrm{y}|\mathrm{x}; \boldsymbol{\theta}) f(\mathrm{x}; \boldsymbol{\theta})$, we obtain the equation

$$\mathrm{L}(\boldsymbol{\theta}) = \mathrm{L_C}(\boldsymbol{\theta}, \mathbf{z}) - \log f(\mathbf{y}|\mathbf{x}; \boldsymbol{\theta}) \qquad [1.4]$$

between the initial log-likelihood $\mathrm{L}(\boldsymbol{\theta})$ and the complete-data log-likelihood $\mathrm{L_C}(\boldsymbol{\theta}, \mathbf{z})$.

### 1.4.2. *Principle*

The EM algorithm is based on the hypothesis that maximizing the complete-data likelihood is simple. Since this likelihood cannot be calculated – y is unknown – an iterative procedure based on the conditional expectation of the log-likelihood for a value of the current parameter $\boldsymbol{\theta}'$ is used as follows: first, calculating the conditional expectation for the two members of equation [1.4], we obtain the fundamental equation of the EM algorithm

$$\mathrm{L}(\boldsymbol{\theta}) = Q(\boldsymbol{\theta}, \boldsymbol{\theta}') - H(\boldsymbol{\theta}, \boldsymbol{\theta}'),$$

where $Q(\boldsymbol{\theta}, \boldsymbol{\theta}') = \mathbb{E}(\mathrm{L_C}(\boldsymbol{\theta}, \mathbf{z})|\mathrm{x}, \boldsymbol{\theta}')$ and $H(\boldsymbol{\theta}, \boldsymbol{\theta}') = \mathbb{E}(\log f(\mathbf{y}|\mathrm{x}; \boldsymbol{\theta})|\mathrm{x}, \boldsymbol{\theta}')$.

Introducing the parameter $\boldsymbol{\theta}'$ allows us to define an iterative algorithm to increase the likelihood. Using Jenssen's inequality, it can be shown that for fixed $\boldsymbol{\theta}'$, the function $H(\boldsymbol{\theta}, \boldsymbol{\theta}')$ is the maximum for $\boldsymbol{\theta} - \boldsymbol{\theta}'$. The value $\boldsymbol{\theta}$ that maximizes $Q(\boldsymbol{\theta}, \boldsymbol{\theta}')$, therefore, satisfies the equation

$$\mathrm{L}(\boldsymbol{\theta}) \geqslant \mathrm{L}(\boldsymbol{\theta}'). \quad [1.5]$$

The EM algorithm involves constructing, from an initial solution $\boldsymbol{\theta}^{(0)}$, the sequence $\boldsymbol{\theta}^{(p)}$ satisfying $\boldsymbol{\theta}^{(q+1)} = \arg\max Q(\boldsymbol{\theta}, \boldsymbol{\theta}^{(q)})$. Equation [1.5] shows that this sequence causes the criterion $\mathrm{L}(\boldsymbol{\theta})$ to develop.

### 1.4.3. *Application to mixture models*

For the mixture model, the complete data are obtained by adding the original component $\mathbf{z}_i$ to each individual member of the sample

$$\mathrm{y} = (\boldsymbol{y}_1, \ldots, \boldsymbol{y}_n) = ((\boldsymbol{x}_1, \boldsymbol{z}_1), \ldots, (\boldsymbol{x}_n, \boldsymbol{z}_n)).$$

Coding $\boldsymbol{z}_i = (z_{i1}, \ldots, z_{ig})$ where, let us recall, $z_{ik}$ equals 1 if $i$ belongs to component $k$ and 0 otherwise, we obtain the following equations

$$f(\mathrm{y}; \boldsymbol{\theta}) = \prod_i f(\boldsymbol{y}_i; \boldsymbol{\theta}) = \prod_i \sum_k \pi_k f(\boldsymbol{x}_i; \boldsymbol{\alpha}_k),$$

$$\mathrm{L_C}(\boldsymbol{\theta}, \mathbf{z}) = \log(f(\mathrm{y}; \boldsymbol{\theta})) = \sum_{i,k} z_{ik} \log\left(\pi_k f(\boldsymbol{x}_i; \boldsymbol{\alpha}_k)\right),$$

$$Q(\boldsymbol{\theta}|\boldsymbol{\theta}') = \sum_{i,k} \mathbb{E}(z_{ik}|\mathrm{x}, \boldsymbol{\theta}') \log\left(\pi_k f(\boldsymbol{x}_i; \boldsymbol{\alpha}_k)\right).$$

Denoting as $\widetilde{z}_{ik}$ the probabilities of belonging $\mathbb{E}(z_{ik}|\mathrm{x}, \boldsymbol{\theta}') = P(z_{ik} = 1|\mathrm{x}, \boldsymbol{\theta}')$, the EM algorithm takes the following form:

– initialize: arbitrarily select an initial solution $\boldsymbol{\theta}$;

– repeat the following two steps until convergence:

- step E (expectation): calculate the probabilities of $x_i$ belonging to the classes, conditionally on the current parameter

$$\widetilde{z}_{ik} = \frac{\pi_k f(\boldsymbol{x}_i, \alpha_k)}{\sum_{k'} \pi_{k'} f(\boldsymbol{x}_i, \alpha_{k'})};$$

- step M (maximization): maximize the log-likelihood conditionally on $\widetilde{z}_{ik}$; the proportions are therefore obtained simply by the equation $\pi_k = \sum_i \widetilde{z}_{ik}/n$, while the parameters $\boldsymbol{\alpha}_k$ are obtained by solving the likelihood equations that depend on the mixture model employed.

### 1.4.4. *Properties*

Under certain conditions of regularity, it has been established that the EM algorithm always converges to a local likelihood maximum. It shows good practical behavior, but may, nevertheless, be quite slow in some situations. This is the case, for instance, when classes are very mixed. This algorithm, proposed by Dempster *et al.* in a seminal paper [DEM 77], often simple to implement, has gained widespread popularity and given rise to a large number of studies that are thoroughly covered in McLachlan and Krishnan's book [MCL 97].

### 1.4.5. *EM: an alternating optimization algorithm*

Hathaway [HAT 86] has shown that the EM algorithm applied to a mixture model may be interpreted as an alternating algorithm for optimizing a fuzzy clustering criterion. We make use of this fact below when we examine the links between estimating the parameters of a mixture model and fuzzy clustering. To obtain this result, Hathaway defines the criterion

$$\mathrm{F_C}(\widetilde{\mathbf{z}}, \boldsymbol{\theta}) = \mathrm{L_C}(\boldsymbol{\theta}, \widetilde{\mathbf{z}}) + \mathrm{H}(\widetilde{\mathbf{z}}), \qquad [1.6]$$

where $\mathrm{H}(\widetilde{\mathbf{z}}) = -\sum_{i,k} \widetilde{z}_{ik} \log \widetilde{z}_{ik}$ is the entropy of the distribution $\widetilde{\mathbf{z}}$. Moreover, if we denote as $\widetilde{\mathbf{z}}_{\theta}$ the posterior distribution $\mathbb{P}(\mathbf{z}|\mathrm{x}, \boldsymbol{\theta})$, it can be shown, using equation [1.4], that the criterion $\mathrm{F_C}$ can also be expressed as

$$\mathrm{F_C}(\widetilde{\mathbf{z}}, \boldsymbol{\theta}) = L(\boldsymbol{\theta}) - \mathrm{KL}(\widetilde{\mathbf{z}}, \widetilde{\mathbf{z}}_{\theta}), \qquad [1.7]$$

where $\mathrm{KL}$ is the Kullback–Liebler divergence between two distributions.

The alternating algorithm for optimizing the criterion $\mathrm{F_C}$, therefore, becomes simple to implement:

– minimizing for fixed $\boldsymbol{\theta}$: equation [1.7] implies that $\widetilde{\mathbf{z}}$ must minimize $\mathrm{KL}(\widetilde{\mathbf{z}}, \widetilde{\mathbf{z}}_{\boldsymbol{\theta}})$ and consequently $\widetilde{\mathbf{z}} = \widetilde{\mathbf{z}}_{\boldsymbol{\theta}}$;

– minimizing for fixed $\widetilde{\mathbf{z}}$: equation [1.6] shows that $\boldsymbol{\theta}$ must maximize the expectation $\mathrm{L_C}(\boldsymbol{\theta}, \widetilde{\mathbf{z}})$.

Therefore, we are dealing with what are precisely the two steps of the $EM$ algorithm. In addition, after each first step, we have $\mathrm{F_C}(\widetilde{\mathbf{z}}, \boldsymbol{\theta}) = \mathrm{F_C}(\widetilde{\mathbf{z}}_{\boldsymbol{\theta}}, \boldsymbol{\theta}) = \mathrm{L}(\boldsymbol{\theta})$, demonstrating that the EM algorithm increases the likelihood.

## 1.5. Clustering and the mixture model

### 1.5.1. *The two approaches*

Mixture models may be used in two different ways to obtain a partition of the initial data.

– The first, known as the *mixture approach*, estimates the parameters of the model and then determines the partition by allocating each individual to the class that maximizes the *a posteriori* probability $\widetilde{\mathbf{z}}_{ik}$ computed using these estimated parameters; this allocation is known as the maximum *a posteriori* probability (MAP) method.

– The second, the *classification approach*, was first presented by Scott and Symons [SCO 71] and developed further by Schroeder [SCH 76]; this approach involves creating a partition of the sample such that each class $k$ is made to correspond to a sub-sample respecting the distribution $f(., \alpha_k)$. This requires simultaneous estimation of the model parameters and the desired partition.

In this section, we describe the criterion that the latter approach optimizes, as well as the optimization algorithm usually employed in this situation. We then briefly compare the two approaches and examine links between these types of methods and the more classical metrical approaches to clustering. We conclude the section by looking at how the mixture model may be interpreted in terms of fuzzy clustering.

### 1.5.2. *Classification likelihood*

Introducing the z partition in the likelihood criterion is not an obvious step, and various ideas have been proposed. Scott and Symons [SCO 71] defined the criterion

$$\mathrm{L_{CR}}(\boldsymbol{\theta}, \mathbf{z}) = \sum_k \sum_{i|z_{ik}=1} \log f(\boldsymbol{x}_i, \alpha_k)$$

in which the proportions do not appear. Symons [SYM 81], realizing that this criterion tends to yield classes of similar proportions, modified it so as to use the complete-data (or classification) log-likelihood described above

$$\mathrm{L_C}(\boldsymbol{\theta}, \mathbf{z}) = \sum_k \sum_{i|z_{ik}=1} \log \pi_k f(\boldsymbol{x}_i, \alpha_k) = \sum_{i,k} z_{ik} \log \pi_k f(\boldsymbol{x}_i, \alpha_k)$$

linked to the previous criterion by the equation

$$\mathrm{L_C}(\boldsymbol{\theta}, \mathbf{z}) = \mathrm{L_{CR}}(\boldsymbol{\theta}, \mathbf{z}) + \sum_k z_{.k} \log \pi_k,$$

where $z_{.k}$ is the cardinal of the class $k$. The quantity $\sum_k z_{.k} \log \pi_k$ is a penalty term that disappears if all the proportions are made to be identical. The criterion $\mathrm{L_{CR}}(\boldsymbol{\theta}, \mathbf{z})$ can, therefore, be seen as a variant of classification likelihood, restricted to a mixture model where all classes have the same proportion.

### 1.5.3. *The CEM algorithm*

When seeking to maximize classification likelihood, it is possible to use a clustering version of the EM algorithm, obtained by adding a clustering step. This yields the very general clustering algorithm known as classification EM (CEM) [CEL 92], defined as follows:

– step 0: arbitrarily select an initial solution $\theta$;

– step E: compute $\widetilde{z}_{ik}$ as in the EM algorithm;

– step C: obtain the z partition by allocating each $x_i$ to the class that maximizes $\widetilde{z}_{ik}$ (MAP); this is equivalent to modifying the $\widetilde{z}_{ik}$ by replacing them with the nearest 1 or 0 values;

– step M: maximize the likelihood depending on the $z_{ik}$; the estimations of the maximum likelihood among the $\pi_k$ and the $\alpha_k$ are obtained using the classes of the partition z as sub-samples, where the proportions are given by the formula $\pi_k = \frac{1}{n} z_{.k}$, the $\alpha_k$ being computed according to the particular mixture model selected.

Here, we have an alternating *dynamic cluster methods* [DID 79] type optimization algorithm, where the $E$ and the $C$ steps correspond to the allocation step, and the $M$ step corresponds to the representation step.

It can be shown that this algorithm is stationary and that it increases the complete-data likelihood at each iteration, given some very general assumptions.

### 1.5.4. *Comparison of the two approaches*

The clustering approach, which determines the parameters at each iteration using truncated mixture model samples, yields a biased and inconsistent estimation, since the number of parameters to be estimated increases as the size of the sample increases. Different authors have studied this problem and shown that it is usually preferable to use the mixture approach.

However, when the classes are well separated and membership relatively small, the clustering approach can sometimes give better results [CEL 93, GOV 96]. Moreover, the CEM algorithm is considerably faster than the EM algorithm, and it may be necessary to use it when computation time is limited, for example in real-time operations, or for very large volumes of data.

Finally, the clustering approach has the advantage of being able to present a large number of clustering algorithms as special cases of the CEM algorithm, which allows it to incorporate them into a probabilistic clustering approach. We will see in section 1.6.3, for example, that the $k$-means algorithm can be seen as a simple special case of the CEM algorithm. In particular, we will show that the optimized criteria, the within-group sum-of-squares criterion for continuous data and the information criterion for qualitative data correspond to the classification likelihood of a particular mixture model. These correspondences, studied in [GOV 89] and [GOV 90b], can be formalized by the following theorem.

THEOREM 1.1.– If the clustering criterion can be expressed as

$$W(\mathbf{z}, \boldsymbol{\lambda}, D) = \sum_{i,k} z_{ik} \Delta(\boldsymbol{x}_i, \boldsymbol{\lambda}_k),$$

where z is a partition of the set to be segmented, $\boldsymbol{\lambda} = (\boldsymbol{\lambda}_1, \ldots, \boldsymbol{\lambda}_g)$ and $\boldsymbol{\lambda}_k$ are representatives of the class $k$, and $\Delta$ is a measure of the dissimilarity between an object $x$ and the representative of a class, and if there exists a real $r$ such that the quantity $\int r^{-\Delta(\boldsymbol{x},\boldsymbol{\lambda})} dx$ is independent of $\boldsymbol{\lambda}$, this criterion is therefore equivalent to the classification likelihood criterion of a mixture model with densities of the form $f(\boldsymbol{x}, \boldsymbol{\lambda}) = \frac{1}{s} r^{-\Delta(\boldsymbol{x}, \boldsymbol{\lambda})}$, $s$ being a positive constant.

This theorem may be used equally well for continuous data as for discrete (binary or qualitative) data – either a Lebesgue measure or a discrete measure will be used accordingly. This theorem is important insofar as a great many clustering criteria can be put into this very general form; for example, it is the case for the intraclass inertia criterion, whose class representative is its center of gravity and where the distance $D$ is the square of the Euclidean distance. It can also help us to fix the fields of application of these criteria and to suggest others.

### 1.5.5. *Fuzzy clustering*

In fuzzy clustering, it is no longer the case that an object either belongs or does not belong to a particular class. Instead, there are degrees of belonging. Formally, fuzzy clustering is characterized by a matrix $\mathrm{c}$ with terms $c_{ik}$ satisfying $c_{ik} \in [0, 1]$ and $\sum_k c_{ik} = 1$. Bezdek's "fuzzy $k$-means" [BEZ 81], one of the most commonly encountered, involves minimizing the criterion

$$\mathrm{W}(\mathbf{c}) = \sum_{i,k} c_{ik}^{\gamma} \mathbf{d}^2(\boldsymbol{x}_i, \mathbf{g}_k),$$

where $\gamma > 1$ is a coefficient for adjusting the degree of fuzziness, $\mathrm{g}_k$ is the center of the class and $\mathrm{d}$ is the Euclidean distance. It is required that $\gamma$ be different from 1, otherwise the function $\mathrm{W}$ is minimal for values of $c_{ik} = 0$ or $1$ and thus we have the usual within-group sum-of-squares criterion. The values usually recommended are between 1 and 2. Minimizing this criterion is achieved using an algorithm that alternates between the two following steps:

1) compute the centers: $g_k = \frac{\sum_i c_{ik}^{\gamma} \boldsymbol{x}_i}{\sum_i c_{ik}}$;

2) compute the fuzzy partition: $c_{ik} = \frac{C_i}{\|\boldsymbol{x}_i - \mathrm{g}_k\|^{\frac{2}{\gamma-1}}}$ with $C_i = \sum_{k'} \frac{1}{\|\boldsymbol{x}_i - \mathrm{g}_{k'}\|^{\frac{2}{\gamma-1}}}$.

Validating this kind of approach and, in particular, choosing the coefficient $\gamma$ can be tricky. Therefore, estimating the parameters of a mixture model is an alternative, more natural, way of addressing this problem. The estimation of the *a posteriori* probabilities $\widetilde{z}_{ik}$ of objects belonging to each class directly provides a fuzzy clustering, and the EM algorithm, applied to the mixture model, may be seen as a fuzzy clustering algorithm.

As mentioned above, Hathaway [HAT 86] went even further and showed that seeking to obtain a fuzzy partition and the parameter $\theta$ using an optimization alternated with a fuzzy clustering criterion leads precisely to the two steps of the EM algorithm, which can therefore be considered as a fuzzy clustering algorithm. One may obtain the same result simply by applying the results from section 1.4.5 to the mixture model. Given that here the probability distribution $\widetilde{\mathrm{z}}$ is defined by the vector $(c_{ik})$ and that we simply have $\mathbb{E}_{\mathbf{c}}(\mathrm{L_C}(\boldsymbol{\theta}, \mathbf{z})) = \mathrm{L_C}(\theta, \mathbf{c})$, we show that the EM algorithm alternately maximizes the criterion

$$\mathrm{W}(\mathbf{c}, \boldsymbol{\theta}) = \mathrm{L_C}(\boldsymbol{\theta}, \mathbf{c}) + \mathrm{H}(\mathbf{c}),$$

where $\mathrm{L_C}$ is the complete-data log-likelihood function where the partition z has been replaced by the fuzzy partition c

$$\mathrm{L_C}(\boldsymbol{\theta}, \mathbf{c}) = \sum_{i,k} c_{ik} \log\left(\pi_k f_k(\boldsymbol{x}_i; \boldsymbol{\alpha})\right)$$

and H is the entropy function

$$\mathrm{H}(\mathbf{c}) = -\sum_{i,k} c_{ik} \log c_{ik}.$$

It is easy to show that if the entropy term of the criterion W is removed, then, "hard" partitions are obtained at each step. The resulting algorithm is simply the CEM algorithm: the difference between the EM and CEM algorithms is the presence of the entropy term. If, when EM converges, the components are highly separated, the fuzzy partition $\mathbf{z}(\boldsymbol{\theta})$ is close to a partition and we have

$$H(\mathbf{z}(\boldsymbol{\theta})) \approx 0$$

and

$$L(\boldsymbol{\theta}) = W(\mathbf{z}(\boldsymbol{\theta}), \boldsymbol{\theta}) = \mathrm{L_C}(\boldsymbol{\theta}, \mathbf{z}(\boldsymbol{\theta})) + H(\mathbf{z}(\boldsymbol{\theta})) \approx \mathrm{L_C}(\boldsymbol{\theta}, \mathbf{z}(\boldsymbol{\theta})).$$

## 1.6. Gaussian mixture model

We will now examine what happens to this approach when each class is modeled by a Gaussian distribution, which is a classical solution for continuous data.

### 1.6.1. *The model*

The pdf of the mixture can be written as $f(\boldsymbol{x}_i; \boldsymbol{\theta}) = \sum_k \pi_k \varphi(\boldsymbol{x}_i; \boldsymbol{\mu}_k, \boldsymbol{\Sigma}_k)$, where $\varphi$ is the pdf of the Gaussian multivariate distribution

$$\varphi(\boldsymbol{x}_i; \boldsymbol{\mu}_k, \boldsymbol{\Sigma}_k) = \frac{1}{(2\pi)^{\frac{d}{2}} |\boldsymbol{\Sigma}_k|^{\frac{1}{2}}} \exp\{-\frac{1}{2}(\boldsymbol{x}_i - \boldsymbol{\mu}_k)^t \boldsymbol{\Sigma}_k^{-1} (\boldsymbol{x}_i - \boldsymbol{\mu}_k)\}$$

and $\boldsymbol{\theta}$ is the vector $(\pi_1, \ldots, \pi_g, \boldsymbol{\mu}_1, \ldots, \boldsymbol{\mu}_g, \boldsymbol{\Sigma}_1, \ldots, \boldsymbol{\Sigma}_g)$ formed by the proportions $\pi_k$ and the parameters $\boldsymbol{\mu}_k$ and $\boldsymbol{\Sigma}_k$, which are, respectively, the mean vector and the covariance matrix of class $k$.

When the sample size is small, or when the dimension of the space is large, the number of parameters must be reduced so as to obtain more parsimonious models. To this end, the spectral decomposition of the matrices [BAN 93, CEL 95] may be used, allowing the covariance matrices to be parameterized uniquely as $\mathbf{\Sigma}_k = \lambda_k \mathbf{D}_k \mathbf{A}_k \mathbf{D}_k^t$, where the diagonal matrix $\mathbf{A}_k$ with determinant 1 and decreasing values defines the *shape* of the class, the orthogonal matrix $\mathbf{D}_k$ defines the *direction* of the class and the positive real number $\lambda_k$ represents the *volume* of the class. Thus, the mixture model is parameterized by the centers $\boldsymbol{\mu}_1, \ldots, \boldsymbol{\mu}_g$, the proportions $\pi_1, \ldots, \pi_g$, the volumes $\lambda_1, \ldots, \lambda_g$, the shapes $\mathbf{A}_1, \ldots, \mathbf{A}_g$ and the directions $\mathbf{D}_1, \ldots, \mathbf{D}_g$ of each class.

For example, when the data are in a plane, $\mathbf{D}$ is a rotation matrix defined by an angle $\alpha$ and $\mathbf{A}$ is a diagonal matrix with diagonal terms $a$ and $1/a$. Figure 1.2 shows the equidensity ellipse of this distribution depending on the values $\alpha$, $\lambda$ and $a$.

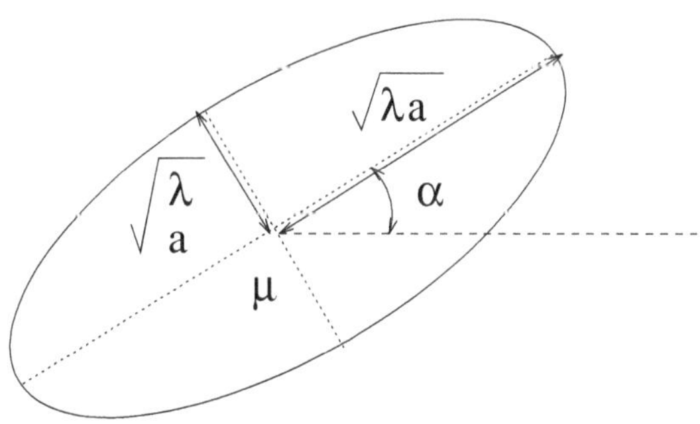


**Figure 1.2.** *Parameterization of a Gaussian class in the plane*

Using this parameterization, it becomes possible to propose solutions that can be seen as a middle way between, on the one hand, restrictive hypotheses (covariance matrices proportional to the identity matrix, or covariance matrices identical for all classes) and, on the other hand, very general constraint-free hypotheses [BAN 93, CEL 95].

This parameterization also highlights two distinct notions that are often conflated under the rather vague heading of *size*: these are, first, the proportion of individuals present within a class and, second, the volume that a class occupies in space. It is quite possible for a class to have a small volume and a high proportion or, alternatively, a large volume but a low proportion.

We will now look at what happens to the CEM algorithm and the classification likelihood criterion in the case of the Gaussian mixture model. It should be noted that a similar approach could be applied to the EM algorithm.

### 1.6.2. *CEM algorithm*

#### 1.6.2.1. *Clustering step*

Each $\boldsymbol{x}_i$ is allocated to the class that maximizes the probability of membership $\widetilde{z}_{ik} = \pi_k \varphi(\boldsymbol{x}_i; \boldsymbol{\mu}_k, \boldsymbol{\Sigma}_k) / (\sum_{k'} \pi_{k'} \varphi(\boldsymbol{x}_i; \boldsymbol{\mu}_{k'}, \boldsymbol{\Sigma}_{k'}))$, that is to say $\pi_k \varphi(\boldsymbol{x}_i; \boldsymbol{\mu}_k, \boldsymbol{\Sigma}_k)$ or, equivalently, the class that minimizes $-\log(\pi_k \varphi(\boldsymbol{x}_i; \boldsymbol{\mu}_k, \boldsymbol{\Sigma}_k))$, which can be written as

$$\mathrm{d}^2_{\boldsymbol{\Sigma}_k^{-1}}(\boldsymbol{x}_i, \boldsymbol{\mu}_k) + \log |\boldsymbol{\Sigma}_k| - 2 \log \pi_k, \qquad [1.8]$$

where $\mathrm{d}^2_{\boldsymbol{\Sigma}_k^{-1}}(\boldsymbol{x}_i, \boldsymbol{\mu}_k)$ is the quadratic distance $(\boldsymbol{x}_i - \boldsymbol{\mu}_k)^t \boldsymbol{\Sigma}_k^{-1} (\boldsymbol{x}_i - \boldsymbol{\mu}_k)$.

1.6.2.2. *Step M*

Here, for a given partition $\mathbf{z}$, we have to determine the parameter $\boldsymbol{\theta}$ that maximizes $\mathrm{L_C}(\boldsymbol{\theta}, \mathbf{z})$, which is equal (ignoring one constant) to

$$-\frac{1}{2}\sum_k \left( \sum_i z_{ik}(\boldsymbol{x}_i - \boldsymbol{\mu}_k)^t \boldsymbol{\Sigma}_k^{-1}(\boldsymbol{x}_i - \boldsymbol{\mu}_k) + z_{.k} \log|\boldsymbol{\Sigma}_k| - 2z_{.k} \log \pi_k \right).$$

The parameter $\boldsymbol{\mu}_k$ is thus necessarily the center of gravity $\overline{\boldsymbol{x}}_k = \frac{1}{z_{.k}} \sum_i z_{ik}\boldsymbol{x}_i$ and the proportions, if they are not constrained, satisfy $\pi_k = z_{.k}/n$. The parameters $\boldsymbol{\Sigma}_k$ must then minimize the function

$$F(\boldsymbol{\Sigma}_1, \ldots, \boldsymbol{\Sigma}_g) = \sum_k z_{.k} \left( \mathrm{trace}(\mathbf{S}_k \boldsymbol{\Sigma}_k^{-1}) + \log|\boldsymbol{\Sigma}_k| \right), \qquad [1.9]$$

where

$$\mathbf{S}_k = \frac{1}{z_{.k}} \sum_i z_{ik}(\boldsymbol{x}_i - \overline{\boldsymbol{x}}_k)(\boldsymbol{x}_i - \overline{\boldsymbol{x}}_k)^t$$

is the covariance matrix of the class $k$. We now examine three particular situations.

### 1.6.3. *Spherical form, identical proportions and volumes*

We now look at the most straightforward situation where all classes have a Gaussian spherical distribution with the same volume and the same proportion. The covariance matrices are written as $\boldsymbol{\Sigma}_k = \lambda \mathbf{D}_k \mathbf{I}_d \mathbf{D}_k^t = \lambda \mathbf{I}_d \ \forall k$, and formula [1.8] shows that individuals can be allotted to the different classes simply by using the usual Euclidean distance $\mathrm{d}^2(\boldsymbol{x}_i, \boldsymbol{\mu}_k)$. Function $F$, therefore, becomes

$$F(\lambda) = \frac{1}{\lambda} \sum_k z_{.k}\, \mathrm{trace}(\mathbf{S}_k) + nd \log \lambda = \frac{1}{\lambda} n\, \mathrm{trace}(\mathbf{S}_W) + nd \log \lambda,$$

where $\mathbf{S}_W = \frac{1}{n}\sum_k z_{.k}\mathbf{S}_k$ is the within-group covariance matrix, thus giving us $\lambda = \frac{\text{trace}(\mathbf{S}_W)}{d}$. The classification likelihood is written as

$$\text{L}_\text{C}(\boldsymbol{\theta}, \mathbf{z}) = -\frac{nd}{2}\log\text{trace}(\mathbf{S}_W) + cst.$$

Maximizing the classification likelihood is therefore equivalent to minimizing the within-group sum-of-squares criterion $\text{trace}(\mathbf{S}_W)$. Moreover, the CEM algorithm is simply the $k$-means algorithm. This means that to use the within-group sum-of-squares criterion is to assume that classes are spherical and have the same proportion and the same volume.

### 1.6.4. *Spherical form, identical proportions but differing volumes*

We now take the model described above and modify it slightly to include classes with different volumes. The covariance matrices are now written $\boldsymbol{\Sigma}_k = \lambda_k\mathbf{I}_p$, and formula [1.8] shows that individuals are allotted to classes according to the distance

$$\frac{1}{\lambda_k}D^2(\boldsymbol{x}_i, \boldsymbol{\mu}_k) + d\log\lambda_k.$$

The distance from a point to the center of a class has been modified by an amount that depends on the volume of the class. This modification has important repercussions; for example, the regions of separation, which in the previous case were hyperplanes, become hyperspheres. It may be shown that the minimized criterion can be written as

$$\sum_k \log\text{trace}(\mathbf{S}_k).$$

With this model, we can very easily recognize situations such as the situation shown in Figure 1.3. Here, the two classes have been simulated with two spherical Gaussian distributions which have the same proportions but widely differing volumes. The result obtained using the classical intraclass inertia criterion corresponds to a separation of the population by a straight line and therefore bears no relation at all to the simulated partition. With the variable-volume model, the obtained partition, shown by the circle, is very close to the initial clustering.

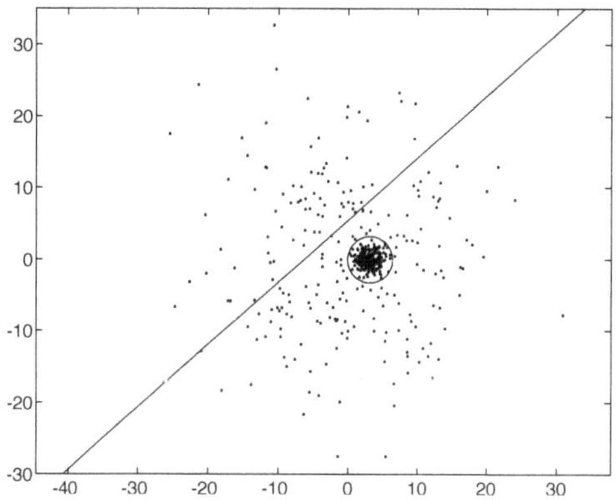


**Figure 1.3.** *Example of classes with different volumes*

It will be noted that without the help of the mixture model, it would have been difficult, on the basis of a simple metrical interpretation, to come up with the distance and the criterion used in this approach.

## 1.6.5. *Identical covariance matrices and proportions*

Our final example is where all classes have the same form and the same proportion. The covariance matrix of each class can thus be written as $\Sigma_k = \Sigma$. It can be shown that individuals are now allotted to classes on the basis of the distance $d^2_{\Sigma^{-1}}(x_i, \mu_k)$ and that the criterion to be minimized may be written as $|\mathbf{S}_W|$, which serves to justify the use of this criterion, sometimes proposed in a metrical context [FRI 67], without reference to the Gaussian model.

## 1.7. Binary data

We now turn, still within a broad discussion of clustering methods based on probability distribution mixture models, to the clustering of sets of individuals measured using binary variables.

### 1.7.1. *Binary mixture model*

As the Gaussian model is often chosen to model each component of the mixture when variables are continuous, the log-linear model [AGR 90, BOC 86] is a natural choice when variables are binary. The complete or saturated log-linear model, where each class has a multinomial distribution with $2^d$ values, is not really applicable in the case of a mixture. Instead, we use a log-linear model with sufficient constraints. The simplest example is the independence model that assumes that, conditionally on membership of a class, the binary variables are independent. From this, we obtain the latent class model [GOO 74, LAZ 68], which we will now examine.

In the binary case, each $x_{ij}$ has a Bernoulli distribution whose probability distribution takes the form

$$f(x_{ij};\alpha_{kj}) = (\alpha_{kj})^{x_{ij}}(1-\alpha_{kj})^{1-x_{ij}} \text{ where } \alpha_{kj} = P(x_{ij}=1|k).$$

Therefore, the distribution of the class $k$ is

$$f_k(\boldsymbol{x}_i;\boldsymbol{\alpha}_k) = \prod_j (\alpha_{kj})^{x_{ij}}(1-\alpha_{kj})^{1-x_{ij}} \text{ where } \boldsymbol{\alpha}_k = (\alpha_{k1},\ldots,\alpha_{kp}),$$

and the mixture model chosen considers that the data $\boldsymbol{x}_1, \ldots, \boldsymbol{x}_n$ constitute a sample of independent instances from a random $\{0, 1\}^d$ probability vector

$$f(\boldsymbol{x}_i; \boldsymbol{\theta}) = \sum_k \pi_k f(\boldsymbol{x}_i; \theta_k) = \sum_k \pi_k \prod_j (\alpha_{kj})^{x_{ij}} (1 - \alpha_{kj})^{1-x_{ij}},$$

where the parameter $\boldsymbol{\theta}$ is constituted by the proportions $\pi_1, \ldots, \pi_g$ and by the parameter vector $\boldsymbol{\alpha} = (\boldsymbol{\alpha}_1, \ldots, \boldsymbol{\alpha}_g)$ of each component.

The problem that arises is how to estimate these parameters, and possibly the origin class as well. As in the case of the Gaussian model, the estimation may be obtained using the EM or the CEM algorithms described above. The only differences concern the computation of the parameters $\alpha_{kj}$ that becomes $\alpha_{kj} = \frac{\sum_i \widetilde{z}_{ik} x_i^j}{\sum_i \widetilde{z}_{ik}}$ for the first, and $\alpha_{kj} = \frac{\sum_i z_{ik} x_i^j}{z_{.k}} =$ % of 1 for the second. Intensive comparisons between the two algorithms EM and CEM were performed by Govaert and Nadif [GOV 96].

### 1.7.2. *Parsimonious model*

The number of parameters of this latent class model is equal to $(g - 1) + g * d$, where $g$, it will be recalled, is the number of classes and $d$ is the number of binary variables. In the case of the complete log-linear model, however, the number of parameters is equal to $2^d$. For example, when $g = 5$ and $d = 10$, the number of parameters for the two models is, respectively, 54 and 1,024. Given that one of the identifiability conditions of the model is that the number of states is greater than the number of parameters, there is a clear interest in being able to propose even more parsimonious models. To this end, the model may be restated as follows

$$f(\boldsymbol{x};\boldsymbol{\theta}) = \sum_k \pi_k \prod_j (\varepsilon_{kj})^{|x_{ij}-a_{kj}|}(1-\varepsilon_{kj})^{1-|x_{ij}-a_{kj}|},$$

where $\begin{cases} a_{kj}=0,\ \varepsilon_{kj}=\alpha_{kj} & \text{if } \alpha_{kj}<0.5 \\ a_{kj}=1,\ \varepsilon_{kj}=1-\alpha_{kj} & \text{otherwise.} \end{cases}$

The parameter $\boldsymbol{\alpha}_k$ is, thus, replaced by the two following parameters:

– a binary vector $\mathbf{a}_k$ representing the center of the class and which is the most frequent binary value for each variable;

– a vector $\boldsymbol{\varepsilon}_k$ belonging to the set $]0,1/2[^d$ that defines the dispersion of the component, and represents the probability of any particular variable's having a value different from that of the center.

We are, thus, led to the parameters used by Aitchinson and Aitken [AIT 76] for discrimination with non-parametric estimation via the kernel method. Starting from this formulation, we arrive at parsimonious situations by stipulating certain constraints: the $[\varepsilon]$ model is defined by stipulating that the dispersion should not depend either on the component or on the variable, the $[\varepsilon_k]$ model by stipulating that it should depend only on the component, and the $[\varepsilon^j]$ model by stipulating that it should depend only on the variable.

For example, in the simplest case, the $[\varepsilon]$ model, given identical proportions ($\pi_k = 1/g$), the clustering approach results in the complete-data log-likelihood being maximized

$$\mathrm{L_C}(\boldsymbol{\theta},\mathbf{z}) = \log\frac{\varepsilon}{1-\varepsilon}\sum_{i,k} z_{ik}\mathbf{d}(\boldsymbol{x}_i,\mathbf{a}_k) + nd\log(1-\varepsilon),$$

which is to say that the criterion

$$W(\mathbf{z}, \boldsymbol{\theta}) = \sum_{i,k} z_{ik} \mathbf{d}(\boldsymbol{x}_i, \mathbf{a}_k) \quad \text{where} \quad \mathbf{d}(\boldsymbol{x}_i, \mathbf{a}_k) = \sum_j |x_{ij} - a_{kj}|$$

is minimized.

Step $E$ of the CEM algorithm, therefore, consists simply of allotting each individual to the class $k$ that minimizes $\mathbf{d}(\boldsymbol{x}_i, \mathbf{a}_k)$. At step $M$, the parameters $a_{kj}$ for each variable $j$ correspond to the majority binary values in each class $k$. A class, therefore, corresponds to a binary vector, and the criterion is easy to interpret: it is simply the number of differences among individuals and their representative in the partition z. To use this binary clustering criterion proposed by different authors [GOW 74, GOV 90a] is therefore to assume that the data come from a particular latent class model.

### 1.7.3. *Examples of application*

To illustrate this approach, we have taken the Stouffer-Toby data set [STO 51], analyzed within the framework of the latent class model by Goodman [GOO 74]. Table 1.1 contains the reactions, classed as one of two possible attitudes, shown by 216 subjects placed in four different conflict situations. We compare the latent class model (here requiring nine parameters) with the log-linear model, with an interaction of order 2, and a very similar number of parameters (11). It is interesting to note that for these data the deviance drops from 7.11 in the case of the linear model to a value of 2.72 in the case of the latent class model. The parameters obtained for the latent class model are shown in Table 1.2.

| $S1$ | $S2$ | $S3$ | $S4$ | Frequencies |
|---|---|---|---|---|
| 1 | 1 | 1 | 1 | 42 |
| 1 | 1 | 1 | 0 | 23 |
| 1 | 1 | 0 | 1 | 6 |
| 1 | 1 | 0 | 0 | 25 |
| 1 | 0 | 1 | 1 | 6 |
| 1 | 0 | 1 | 0 | 24 |
| 1 | 0 | 0 | 1 | 7 |
| 1 | 0 | 0 | 0 | 38 |
| 0 | 1 | 1 | 1 | 1 |
| 0 | 1 | 1 | 0 | 4 |
| 0 | 1 | 0 | 1 | 1 |
| 0 | 1 | 0 | 0 | 6 |
| 0 | 0 | 1 | 1 | 2 |
| 0 | 0 | 1 | 0 | 9 |
| 0 | 0 | 0 | 1 | 2 |
| 0 | 0 | 0 | 0 | 20 |

**Table 1.1.** *Stouffer-Toby data*

| – | $p_k$ | $a_{k1}$ | $a_{k2}$ | $a_{k3}$ | $a_{k4}$ |
|---|---|---|---|---|---|
| 1 | 0.279 | 0.993 | 0.940 | 0.927 | 0.769 |
| 2 | 0.721 | 0.714 | 0.330 | 0.354 | 0.132 |

**Table 1.2.** *Results obtained by EM for the latent class model*

## 1.8. Categorical variables

We now extend the results of the previous section to categorical data.

### 1.8.1. *Multinomial mixture model*

As for binary data, we will examine the latent class model and therefore assume that the $d$ qualitative variables are independent, conditionally on their membership to a class. If

$\alpha_{kj}^h$ is the probability that the $j$th variable takes the modality $h$ when an individual belongs to the class $k$, therefore the pdf of the mixture can be written as

$$f(\boldsymbol{x}_i;\boldsymbol{\theta}) = \sum_k \pi_k f(\boldsymbol{x}_i;\boldsymbol{\alpha}_k) = \sum_k \pi_k \prod_j \prod_{h=1}^{m_j} (\alpha_{kj}^h)^{x_{ij}^h},$$

where the parameter $\boldsymbol{\theta}$ is defined by the proportions $\pi_1,\ldots,\pi_g$ and by the parameters $\boldsymbol{\alpha}_k = (\alpha_{kj}^h; j = 1,\ldots,d; h = 1,\ldots,m_j)$ of the pdf of each component.

As before, estimating the parameter $\boldsymbol{\theta}$, and possibly estimating the native class of each of the $\boldsymbol{x}_i$, may be achieved by maximizing the likelihood $\mathrm{L}(\boldsymbol{\theta};\mathrm{x})$ using the EM algorithm, or by maximizing the complete-data likelihood $\mathrm{L_C}(\boldsymbol{\theta},\mathbf{z})$ using the CEM algorithm. In the case of the EM algorithm, the computation of the parameters $\boldsymbol{\alpha}_k$ at step M is defined by the equation $\alpha_{kj}^h = \sum_i \widetilde{z}_{ik} x_{ij}^h / \sum_i \widetilde{z}_{ik}$, where $\widetilde{z}_{ik}$ are the probabilities obtained in the usual fashion at step E. In the case of the CEM algorithm, the computation of the parameters $\boldsymbol{\alpha}_k$ becomes $\alpha_{kj}^h = \sum_i z_{ik} x_{ij}^h / z_{.k}$, where $\mathbf{z}$ is the partition obtained by the MAP from the probabilities $\widetilde{z}_{ik}$.

We now look at what happens to the complete-data likelihood criterion when the clustering approach is used with the assumption that the proportions $\pi_k$ are constant. If we denote $s_{ij}^k = \sum_i z_{ik} x_{ij}^h$, $s_{.}^{jh} = \sum_i x_{ij}^h$, $s_k^{..} = \sum_j \sum_{h=1}^{m_j} s_{ij}^k$ and $s_{.}^{..} = \sum_{i,j} \sum_{h=1}^{m_j} x_{ij}^h = nd$, we can easily show that the equation

$$\mathrm{L_C}(\mathbf{z},\boldsymbol{\theta}) = \sum_{k,j} \sum_{h=1}^{m_j} s_{ij}^k \log \alpha_{kj}^h$$

is obtained.

Given that, at convergence, $\alpha_{kj}^{h} = s_{ij}^{k}/z_{.k}$, it can be shown [CEL 91] that the CEM algorithm maximizes the information criterion [BEN 73a]

$$H(\mathbf{z}, J) = \sum_{k,j} \sum_{h=1}^{m_j} \frac{s_{ij}^{k}}{s_{\cdot}^{\cdot\cdot}} \log \frac{s_{\cdot}^{\cdot\cdot}\, s_{k}^{jk}}{s_{k}^{\cdot\cdot} s_{\cdot}^{jh}}$$

which represents the information from the initial table, retained by the partition z, and yielding results very close to the $\chi^2$ criterion

$$\chi^2(\mathbf{z}, J) = \sum_{k,j} \sum_{h=1}^{m_j} \frac{(s_{\cdot}^{\cdot\cdot}\, s_{ij}^{k} - s_{k}^{\cdot\cdot} s_{\cdot}^{jh})^2}{s_{\cdot}^{\cdot\cdot}\, s_{k}^{\cdot\cdot} s_{\cdot}^{jh}}.$$

Therefore, it follows that to seek a partition into $g$ classes maximizing the information criterion or the $\chi^2$ criterion (approximately equivalent) is to assume that the data derive from a latent class model.

A parallel may be drawn here with the analysis of multiple correspondences. It is not difficult to see that with the geometrical representation used in this factorial analysis, the $\chi^2$ criterion is quite simply the familiar criterion of intraclass inertia.

### 1.8.2. *Parsimonious model*

The number of parameters $(g - 1) + g * \sum_j (m_j - 1)$ required by the latent class model that we have just described is usually considerably smaller than the number of parameters $\prod_j m_j$ required by the complete log-linear model. For example, for a number of classes $g$ is equal to 5 and a number of qualitative variables $d$ are equal to 10, and where the number of modalities $m_j$ is 4 for all the variables, the number of parameters for the two models is, respectively, 154 and $10^6$. In many cases, this number will be quite excessive, and more parsimonious models are called for.

To this end, we begin by remarking that if, for each variable $j$, the modality of highest probability is denoted as $h^*$, the model may therefore be re-parameterized as follows

$$f(\boldsymbol{x}_i;\boldsymbol{\theta}) = \sum_k \pi_k \prod_j \left( (1-\varepsilon_{kj}^{h^*})^{1-d(x_{ij},a_{kj})} \prod_{h \neq h^*} (\varepsilon_{kj}^{h})^{|x_{ij}^h - a_k^{jh}|} \right),$$

where $\mathbf{a}_{kj} = (a_{kj}^1,\ldots,a_{kj}^{m_j})$ with $a_{kj}^h = 1$ if $h = h^*$ and 0 otherwise, $\boldsymbol{\varepsilon}_{kj} = (\varepsilon_{kj}^1,\ldots,\varepsilon_{kj}^{m_j})$ where $\varepsilon_{kj}^h = 1-\alpha_{kj}^h$ if $h = h^*$ and $\alpha_{kj}^h$ otherwise, and $\delta(x_{ij},a_{kj}) = 0$ if $x_{ij}$ and $a_{kj}$ take the same modality and 1 otherwise. Like for binary data, the vector $\mathbf{a}_k = (\mathbf{a}_{k1},\ldots,\mathbf{a}_{kd})$ may be interpreted as the center of the class $k$ and the vectors $\varepsilon_k^j$ as dispersions. For example, if the parameter $\alpha_{kj}$ is equal to the vector $(0.7, 0.2, 0.1)$, the new parameters become $\mathbf{a}_{kj} = (1,0,0)$ and $\boldsymbol{\varepsilon}_{kj} = (0.3, 0.2, 0.1)$.

With the model restated in this way, it is possible to introduce simple constraints, such as requiring non-majority modalities to have the same dispersion

$$\begin{cases} \varepsilon_{kj}^{h^*} = \varepsilon_{kj} \\ \varepsilon_{kj}^{h} = (1-\varepsilon_{kj})/(m_j-1) \text{ for } h \neq h^*. \end{cases}$$

This gives us a model used in discrimination, where the number of parameters has been reduced from $(g-1) + g * \sum_j (m_j - 1)$ to $(g-1) + \sum_j (m_j-1)$.

Even more parsimonious models may be obtained if additional constraints are placed on dispersions, for example by requiring that $\varepsilon_{kj}^{h^*}$ should not depend on the variable (model $[\varepsilon_k]$), on the class (model $[\varepsilon_j]$) or on neither the variable nor the class (model $[\varepsilon]$). If we restrict ourselves to this last model $[\varepsilon]$, and if we require proportions to be equal, the complete-data log-likelihood can be expressed quite simply

$$\mathrm{L_C}(\mathbf{z}, \boldsymbol{\theta}) = \log \frac{\varepsilon}{1-\varepsilon} \sum_k \sum_{i=1}^{n} z_{ik} \mathbf{d}(\boldsymbol{x}_i, \mathbf{a}_k) + nd \log(1-\varepsilon),$$

where $\mathrm{d}(\boldsymbol{x}_i, \mathrm{a}_k)$ is a distance reflecting the number of different modalities between the vector $\boldsymbol{x}_i$ and the center $\mathrm{a}_k$. At the clustering step of the CEM algorithm, the individuals $i$ are thus allotted to the class $k$ that minimizes $\mathrm{d}(\boldsymbol{x}_i, \mathrm{a}_k)$, and at step M, the co-ordinates $a_{kj}$ of the centers $\mathrm{a}_k$ are obtained by taking the majority of modalities. Furthermore, Jollois and Nadif [JOL 02] considered the clustering of categorical data under the classification maximum likelihood approach. In this setting, with a parsimonious multinomial mixture model, they defined a generalization of the $k$-modes criterion [HUA 98]. They showed that $k$-modes is just a particular version of CEM and the $k$-modes criterion is associated with a multinomial mixture model $[\varepsilon]$ with supplementary constraints that are too restrictive: the proportions are assumed to be equal and the variables to have the same number of categories. They conducted experiments showing the superiority of CEM with the model on $k$-modes when these assumptions are not verified.

In practice, we suggest taking very simple models by class, for example latent class models with a single parameter, and then increasing the number of components if necessary.

When the qualitative variables are ordinal, it is possible either to convert the data into binary data or to use an approach similar to the approach we have just described, taking into account the order that exists among the modalities.

## 1.9. Contingency tables

To measure the information provided by a contingency table, we need to evaluate the links existing between the two sets $I$ and $J$ as discussed in the Introduction. Several measures of association exist, and one of the most frequently employed is the phi-squared criterion $\phi^2$ (see Introduction and Chapter 4). This criterion, used, for example, in correspondence analysis (CA), is defined as follows

$$\phi^2(I, J) = \sum_{i,j} \frac{(p_{ij} - p_{i.}p_{.j})^2}{p_{i.}p_{.j}}.$$

The phi-squared criterion can be used to evaluate the quality of a partition z of $I$: to this end, we associate the partition z with the phi-squared $\phi^2(\mathbf{z}, J)$ of the contingency table with $g$ rows and $d$ columns obtained from the initial table in computing the sum of the rows of each cluster. It can be shown that

$$\phi^2(I, J) \geq \phi^2(\mathbf{z}, J) \qquad [1.10]$$

and therefore the proposed regrouping necessarily leads to a loss of information. The objective of classification is to find the partition z that minimizes this loss, i.e. which maximizes $\phi^2(\mathbf{z}, J)$. We notice that when the row profiles are equal for each cluster, inequality [1.10] becomes $\phi^2(\mathbf{z}, J) = \phi^2(I, J)$, and in this particular case, there is no loss of information. In addition, the problem is meaningful only when the number of clusters is fixed. Otherwise, the optimal partition is simply the partition where each element of $I$ forms a cluster.

### 1.9.1. *MNDKI2 algorithm*

The MNDKI2 algorithm is based on the same geometrical representation of a contingency table as that used in CA. This representation is justified for several reasons, in particular, because of the similar role played by each of the two dimensions in the analyzed table, and also because of the property of distributional equivalence, which implies stable results when agglomerating elements with similar profiles. In this representation, each row $i$ corresponds to a point vector $\mathbb{R}^d$ defined by the profile $p_J^i$ weighted by the marginal frequency $p_{i.}$. The distances among profiles is not defined by the usual Euclidean metric, but instead by the weighted Euclidean metric, known as the *chi-squared metric* $D^2$, defined by the diagonal matrix $\text{diag}(\frac{1}{p_{.1}}, \ldots, \frac{1}{p_{.d}})$.

If $\mathbf{z}$ is a partition of the rows, we can define the frequencies $p_{kj} = \sum_i z_{ik} p_{ij}$ and the average row profile of the $k$th cluster

$$p_J^k = \left(\frac{p_{k1}}{p_{k.}}, \ldots, \frac{p_{kd}}{p_{k.}}\right)^t,$$

where $p_{k.} = \sum_j p_{kj}$. With this representation, we can show after some calculation that the total of squared distances $T$, the between-cluster sums of squares $B(\mathbf{z})$ and the within-cluster sums of squares $W(\mathbf{z})$ can be written as

$$T = \sum_i p_{i.} D^2(p_J^i, p_J) = \phi^2(I, J),$$

$$B(\mathbf{z}) = \sum_k p_{k.} D^2(p_J^k, p_J) = \phi^2(\mathbf{z}, J),$$

and

$$\mathrm{W}(\mathbf{z}) = \sum_{i,k} z_{ik} p_{i.} D^2(p_J^i, p_J^k).$$

The traditional equation between the total of squared distances, the within-cluster sums of squares and the between-cluster sums of squares $T = \mathrm{W}(\mathbf{z}) + B(\mathbf{z})$ leads to the following equation

$$\phi^2(I, J) = \mathrm{W}(\mathbf{z}) + \phi^2(\mathbf{z}, J).$$

The term $\mathrm{W}(\mathbf{z})$, therefore, represents the information lost when grouping the elements according to the partition z, and $\phi^2(\mathbf{z}, J)$ corresponds to the information which is preserved. Consequently, since the quantity $\phi^2(I, J)$ does not depend on the partition z, looking for the partition maximizing the criterion $\phi^2(\mathbf{z}, J)$ is equivalent to looking for the partition minimizing criterion $W(\mathbf{z})$. To minimize this criterion, it is possible to apply $k$-means to the set of profiles with the $\chi^2$ metric. An iterative algorithm, known as MNDKI2, is thus obtained, locally maximizing $\phi^2(\mathbf{z}, J)$.

The question that naturally arises is: which probabilistic model does the criterion $\phi^2(\mathbf{z}, J)$ minimized by the MNDKI2 algorithm correspond to? The answer to this question will not only shed some light on this criterion, but it will also help us to propose other criteria. This is a question that we will focus on. Unfortunately, unlike the standard $k$-means algorithm, the MNDKI2 algorithm does not correspond to the classification approach associated with a mixture model [GOV 89]. However, using a mixture of multinomial distributions, examined in the following section, we will obtain approximately similar properties.

### 1.9.2. *Model-based approach*

#### 1.9.2.1. *Multinomial mixture*

A contingency table can be obtained using a mixture of multinomial distributions by the following process of simulation [GOV 07]:

– z: each individual is allotted to a class according to a multinomial distribution with parameters $(\pi_1, \ldots, \pi_g)$;

– $\mathrm{x}_I = (x_{1.}, \ldots, x_{n.})$: generate each row sum $x_{i.}$ according to a discrete distribution $\psi$ such as a Poisson or a binomial distribution;

– x: each $x_i$ is assumed to arise from a multinomial distribution with parameters $x_{i.}$ and $\alpha_{k1}, \ldots, \alpha_{kd}$.

Thus, if $\boldsymbol{\theta} = (\pi_1, \ldots, \pi_g, \alpha_{11}, \ldots, \alpha_{gd})$ denotes the parameter of the model and $\varphi$ is the multinomial distribution of the $k$th component, the pdf of this model is written as

$$\begin{aligned} f(\boldsymbol{x}_i; \boldsymbol{\theta}) &= \psi(x_{i.}) \sum_k \pi_k \varphi(\boldsymbol{x}_i; x_{i.}, \alpha_{k1}, \ldots, \alpha_{kd}) \\ &= \psi(x_{i.}) \sum_k \pi_k \frac{x_{i.}!}{x_{i1}! \ldots x_{is}!} \alpha_{k1}^{x_{i1}} \ldots \alpha_{kd}^{x_{id}} \\ &= A \sum_k \pi_k \alpha_{k1}^{x_{i1}} \ldots \alpha_{kd}^{x_{id}}, \end{aligned}$$

where $A = \psi(x_{i.}) \frac{x_{i.}!}{x_{i1}! \ldots x_{id}!}$ does not depend on the parameter $\boldsymbol{\theta}$. The log-likelihood (without the additional constant $\log A$) can therefore be written as

$$L(\boldsymbol{\theta}; \mathrm{x}) = \sum_i \log \sum_k \pi_k \alpha_{k1}^{x_{i1}} \ldots \alpha_{kd}^{x_{id}}, \qquad [1.11]$$

and the complete data log-likelihood is as follows

$$L(\boldsymbol{\theta}; \mathrm{x}, \mathrm{z}) = \sum_{i,k} z_{ik} \left( \ln \pi_k + \sum_j x_{ij} \log \alpha_{kj} \right). \qquad [1.12]$$

The classical problem is, therefore, to estimate the parameter $\theta$ from the sample. In the clustering context, the mixture model serves to find the component from which each row arises. Next, we see how the EM and CEM algorithms allow us to achieve this goal.

### 1.9.2.2. *EM algorithm*

For this multinomial mixture model, the application of the EM algorithm described in section 1.4.3 to the sample $\mathrm{x} = (x_1, \ldots, x_n)$ leads in the M-step to $\alpha_{kj} = \frac{\sum_i \tilde{z}_{ik} x_{ij}}{\sum_i \tilde{z}_{ik} x_{i.}}$. The different steps of EM are then expressed in algorithm 1.1.

**Algorithm 1.1** Multinomial EM

**input:** x, $g$
**initialization:** z, $\pi_k = \frac{z_{.k}}{n}$, $\alpha_{kj} = \frac{\sum_i z_{ik} x_{ij}}{\sum_i z_{ik} x_{i.}}$
**repeat**
  **E-step.** $\tilde{z}_{ik} \propto \pi_k \alpha_{k1}^{x_{i1}} \ldots \alpha_{kd}^{x_{id}}$
  **M-step.** $\pi_k = \frac{\tilde{z}_{.k}}{n}$, $\alpha_{kj} = \frac{\sum_i \tilde{z}_{ik} x_{ij}}{\sum_i \tilde{z}_{ik} x_{i.}}$
**until** convergence
**return** $\pi, \alpha$

In the maximum likelihood approach of the classical mixture model, after we have estimated the parameter $\theta$, we can give a probabilistic clustering of the $n$ rows in terms of their fitted posterior probabilities of component membership, and obtain a partition using a classification step that assigns each object to the component of the mixture to which it has the highest posterior probability of belonging.

### 1.9.2.3. *CEM algorithm*

Recall that in this classification approach, a C-step that converts the posterior probabilities $\tilde{z}_{ik}$s to a discrete classification is included prior to performing the M-step. The different steps of the CEM algorithm are then expressed in algorithm 1.2.

---

**Algorithm 1.2** Multinomial CEM

**input:** x, $g$
**initialization:** z, $\pi_k = \frac{z_{.k}}{n}$, $\alpha_{kj} = \frac{\sum_i z_{ik} x_{ij}}{\sum_i z_{ik} x_{i.}}$
**repeat**
  **E-step.** $\widetilde{z}_{ik} \propto \pi_k \alpha_{k1}^{x_{i1}} \ldots \alpha_{kd}^{x_{id}}$
  **C-step.** $z_i = \arg\max_k \widetilde{z}_{ik}$
  **M-step.** $\pi_k = \frac{z_{.k}}{n}$, $\alpha_{kj} = \frac{\sum_i z_{ik} x_{ij}}{\sum_i z_{ik} x_{i.}}$
**until** convergence
**return** $\pi$, $\alpha$, z,

---

Having established an estimate of the parameters, and denoting $p_{kj} = \frac{x_{kj}}{x_{k.}}$, we can express the criterion as

$$L(\boldsymbol{\theta}; \mathbf{x}, \mathbf{z}) = \sum_k z_{.k} \ln \pi_k + \sum_{k,j} x_{kj} \log \frac{x_{kj}}{x_{k.}}$$
$$= \sum_k z_{.k} \ln \pi_k + n \sum_{k,j} p_{kj} \log \frac{p_{kj}}{p_{k.} p_{.j}} + n \sum_j p_{.j} \log p_{.j}. \quad [1.13]$$

Note that the term $\sum_{k,j} p_{kj} \log \frac{p_{kj}}{p_{k.} p_{.j}}$ is the mutual information $\mathcal{I}(\mathbf{z}, J)$ quantifying the information shared between z and $J$. This can easily be shown using the definition in terms of entropies

$$\mathcal{I}(\mathbf{z}, J) = \mathrm{H}(\mathbf{z}) + \mathrm{H}(J) - \mathrm{H}(\mathbf{z}, J),$$

where H(.) is the entropy. This mutual information can be linked to the $\phi^2$ criterion as follows: first, using the equalities $\sum_{k,j} p_{k.} p_{.j} = 1$ and $\sum_{k,j} p_{kj} = 1$, we have the equation

$$\sum_{k,j} \frac{(p_{kj} - p_{k.} p_{.j})^2}{p_{k.} p_{.j}} = \sum_{k,j} p_{k.} p_{.j} \left( \left( \frac{p_{kj}}{p_{k.} p_{.j}} \right)^2 - 1 \right).$$

Second, using the approximation $x^2 - 1 \approx 2x \log x$, excellent in the neighborhood of 1 and good in the interval $[0, 3]$, the approximation

$$\sum_{k,j} p_{k.}p_{.j}\left(\left(\frac{p_{kj}}{p_{k.}p_{j.}}\right)^2 - 1\right) \approx 2\sum_{k,j} p_{kj} \log \frac{p_{kj}}{p_{k.}p_{.j}}$$

can be obtained [BEN 73b]. Finally, we have the following approximation

$$\sum_{k,j} p_{kj} \log \frac{p_{kj}}{p_{k.}p_{.j}} \approx \frac{1}{2}\sum_{k,j} \frac{(p_{kj} - p_{k.}p_{.j})^2}{p_{k.}p_{.j}}, \qquad [1.14]$$

and therefore,

$$\mathcal{I}(\mathbf{z}, J) \approx \frac{1}{2}\phi^2(\mathbf{z}, J).$$

Therefore, from equations [1.13] and [1.14], when the proportions are fixed, the maximization of $\mathrm{L_C}(\boldsymbol{\theta}; \mathbf{z})$ is equivalent to the maximization of the mutual information $\mathcal{I}(\mathbf{z}, J)$, and approximately equivalent to the maximization of the phi-squared criterion $\phi^2(\mathbf{z}, J)$: the use of the two criteria $\phi^2(\mathbf{z}, J)$ and $I(\mathbf{z}, J)$, therefore, is based on the implicit assumption that the data arise from a mixture of multinomial distributions.

### 1.9.3. *Illustration*

To illustrate the results obtained by MNDKI2, we will use a CA representation. Let us recall that CA is an exploratory multivariate technique that converts a contingency table into a particular type of graphical display in which the rows and the columns of the matrix are depicted as points [BEN 73b, GRE 88b, LEB 84]. It can be used on any two-way table, sparse or not, as the case may be. It projects the rows and columns of a data matrix into points within a graph in a

Euclidean space. The graph is, therefore, used to gain some understanding of the data and to extract information from it. To show the links between MNDKI2 and CA, we will use a comparison of time-budgets as an example [JAM 76]. We have a data matrix where $x_{ij}$ represents the amount of time spent on a variety of activities $i$ by a population $j$ during a given time period. The set $I$ comprises 10 activity clusters: prof (professional), tran (transport), home (housework), child (activities pertaining to childcare), shop (shopping), wash (washing and personal care), meal (mealtime), sleep, tv (television) and leis (other leisure activities). The set $J$ is composed of 28 types of population characterized by gender, country, professional activity and marital status. The vector of letters identifying each population can be interpreted as follows: m or w (man or woman), a or na (active or not active professionally), s or ns (single or not single), us, we, ea or yu (USA, western country, eastern country or Yugoslavia); for instance, *mnsyu* corresponds to a man, not single and from Yugoslavia. We have presented the data in Table 1.3.

Here, we present the best result obtained by MNDKI2 from among 10 random initial positions when the number of clusters in partition z is 3. The initial $\phi^2(I, J)$ value is 9658.38 and the resulting $\phi^2(\mathrm{z}, J)$ value is 8386.83. The percentage of $\phi^2$ accounted for by the partition is very good in this small example: more than 86% of the $\phi^2$ is preserved. The clusters in the obtained partition z are the following: cluster 1: home, child; cluster 2: prof, tran; and cluster 3: sleep, wash, leis, meal, shop, tv.

The column profiles $\frac{p_{ij}}{p_{i.}p_{.j}}$ (with a multiple coefficient $f_{i.}$) reorganized according to z are reported in Table 1.4. We observe the similarity of the profiles belonging to each cluster. The most interesting values are those that are a long way from the mean 1. They characterize the partition: for example, the category $wnaus$ is a characteristic of the clusters 1 and 2.

| | prof | tran | home | child | shop | wash | meal | sleep | tv | leis |
|---|---|---|---|---|---|---|---|---|---|---|
| maus | 610 | 140 | 60 | 10 | 120 | 95 | 115 | 760 | 175 | 315 |
| waus | 475 | 90 | 250 | 30 | 140 | 120 | 100 | 775 | 115 | 305 |
| wnaus | 10 | 0 | 495 | 110 | 170 | 110 | 130 | 785 | 160 | 430 |
| mnsus | 615 | 141 | 65 | 10 | 115 | 90 | 115 | 765 | 180 | 305 |
| wnsus | 179 | 29 | 421 | 87 | 161 | 112 | 119 | 776 | 143 | 373 |
| msus | 585 | 115 | 50 | 0 | 150 | 105 | 100 | 760 | 150 | 385 |
| wsus | 482 | 94 | 196 | 18 | 141 | 130 | 96 | 775 | 132 | 336 |
| mawe | 652 | 100 | 95 | 7 | 57 | 85 | 150 | 807 | 115 | 330 |
| wawe | 510 | 70 | 307 | 30 | 80 | 95 | 142 | 815 | 87 | 262 |
| wnawe | 20 | 7 | 567 | 87 | 112 | 90 | 180 | 842 | 125 | 367 |
| mnswe | 655 | 97 | 97 | 10 | 52 | 85 | 152 | 807 | 122 | 320 |
| wnswe | 168 | 22 | 529 | 69 | 102 | 83 | 174 | 825 | 119 | 392 |
| mswe | 642 | 105 | 72 | 0 | 62 | 77 | 140 | 812 | 100 | 387 |
| wswe | 389 | 34 | 262 | 14 | 92 | 97 | 147 | 848 | 84 | 392 |
| mayu | 650 | 140 | 120 | 15 | 85 | 90 | 105 | 760 | 70 | 365 |
| wayu | 560 | 105 | 375 | 45 | 90 | 90 | 95 | 745 | 60 | 235 |
| wnayu | 10 | 10 | 710 | 55 | 145 | 85 | 130 | 815 | 60 | 380 |
| mnsyu | 650 | 145 | 112 | 15 | 85 | 90 | 105 | 760 | 80 | 357 |
| wnsyu | 260 | 52 | 576 | 59 | 116 | 85 | 117 | 775 | 65 | 295 |
| msyu | 615 | 125 | 95 | 0 | 115 | 90 | 85 | 760 | 40 | 475 |
| wsyu | 413 | 89 | 318 | 23 | 112 | 96 | 102 | 774 | 45 | 409 |
| maea | 650 | 142 | 122 | 22 | 76 | 94 | 100 | 764 | 96 | 334 |
| waea | 578 | 106 | 338 | 42 | 106 | 94 | 52 | 752 | 64 | 228 |
| wnaea | 24 | 8 | 594 | 72 | 158 | 92 | 128 | 840 | 86 | 398 |
| mnsea | 652 | 133 | 134 | 22 | 68 | 94 | 102 | 762 | 122 | 310 |
| wnsea | 434 | 77 | 431 | 60 | 117 | 88 | 105 | 770 | 73 | 229 |
| msea | 627 | 148 | 68 | 0 | 88 | 92 | 86 | 770 | 58 | 463 |
| wsea | 433 | 86 | 296 | 21 | 128 | 102 | 94 | 758 | 58 | 379 |

**Table 1.3.** *Transposed time-budget data matrix*

To illustrate the relationship between CA and MNDKI2, we have shown in Figure 1.4 the representation of $I$ on the first two axes that account for 84% of $\phi^2$. We can observe that clusters 1 and 2 are strongly opposed and cluster 3 is the middle cluster.

## 1.10. Implementation

There are a number of software developments implementing the methods described in this chapter, not least among them is the MIXMOD[1] program. In this section, we give a quick overview of the problems that software implementations need to address.

| | prof | tran | home | child | shop | wash | meal | sleep | tv | leis |
|---|---|---|---|---|---|---|---|---|---|---|
| maus | 1359 | 1624 | 216 | 300 | 1103 | 1000 | 985 | 968 | 1758 | 903 |
| waus | 1058 | 1044 | 901 | 899 | 1286 | 1263 | 856 | 987 | 1155 | 874 |
| wnaus | 22 | 0 | 1785 | 3297 | 1562 | 1158 | 1113 | 1000 | 1607 | 1232 |
| mnsus | 1370 | 1635 | 234 | 300 | 1056 | 947 | 984 | 974 | 1807 | 874 |
| wnsus | 399 | 336 | 1518 | 2607 | 1479 | 1179 | 1019 | 988 | 1436 | 1069 |
| msus | 1304 | 1334 | 180 | 0 | 1378 | 1105 | 856 | 968 | 1507 | 1103 |
| wsus | 1074 | 1091 | 707 | 539 | 1296 | 1369 | 822 | 987 | 1326 | 963 |
| mawe | 1454 | 1161 | 343 | 210 | 524 | 896 | 1285 | 1029 | 1156 | 947 |
| wawe | 1137 | 813 | 1108 | 900 | 736 | 1001 | 1217 | 1039 | 875 | 752 |
| wnawe | 45 | 81 | 2047 | 2611 | 1030 | 949 | 1543 | 1074 | 1257 | 1053 |
| mnswe | 1461 | 1127 | 350 | 300 | 478 | 896 | 1303 | 1029 | 1227 | 918 |
| wnswe | 362 | 247 | 1844 | 1999 | 906 | 845 | 1440 | 1015 | 1155 | 1086 |
| mswe | 1432 | 1220 | 260 | 0 | 570 | 812 | 1200 | 1035 | 1006 | 1111 |
| wswe | 882 | 401 | 961 | 427 | 860 | 1039 | 1280 | 1099 | 858 | 1143 |
| mayu | 1448 | 1624 | 433 | 450 | 781 | 947 | 899 | 968 | 703 | 1046 |
| wayu | 1248 | 1218 | 1352 | 1349 | 827 | 947 | 813 | 949 | 603 | 674 |
| wnayu | 22 | 116 | 2560 | 1648 | 1332 | 895 | 1113 | 1038 | 603 | 1089 |
| mnsyu | 1449 | 1683 | 404 | 450 | 781 | 948 | 899 | 968 | 804 | 1024 |
| wnsyu | 579 | 603 | 2077 | 1768 | 1066 | 895 | 1002 | 987 | 653 | 845 |
| msyu | 1370 | 1450 | 343 | 0 | 1057 | 947 | 728 | 968 | 402 | 1361 |
| wsyu | 928 | 1041 | 1156 | 695 | 1037 | 1019 | 880 | 994 | 456 | 1182 |
| maea | 1448 | 1648 | 440 | 659 | 698 | 990 | 856 | 973 | 964 | 957 |
| waea | 1310 | 1251 | 1239 | 1280 | 991 | 1006 | 453 | 974 | 654 | 665 |
| wnaea | 53 | 93 | 2142 | 2158 | 1452 | 969 | 1096 | 1070 | 864 | 1141 |
| mnsea | 1454 | 1544 | 483 | 660 | 625 | 990 | 874 | 971 | 1226 | 889 |
| wnsea | 974 | 899 | 1564 | 1810 | 1082 | 933 | 905 | 987 | 738 | 661 |
| mnsea | 1397 | 1717 | 245 | 0 | 809 | 969 | 736 | 981 | 583 | 1327 |
| wsea | 983 | 1017 | 1088 | 641 | 1199 | 1094 | 820 | 984 | 594 | 1107 |

**Table 1.4.** *The* $1,000 \times \frac{p_{ij}}{p_{i.}p_{.j}}$ *profiles reorganized according to the partition* $\mathbf{z}$

1 www.mixmod.org.

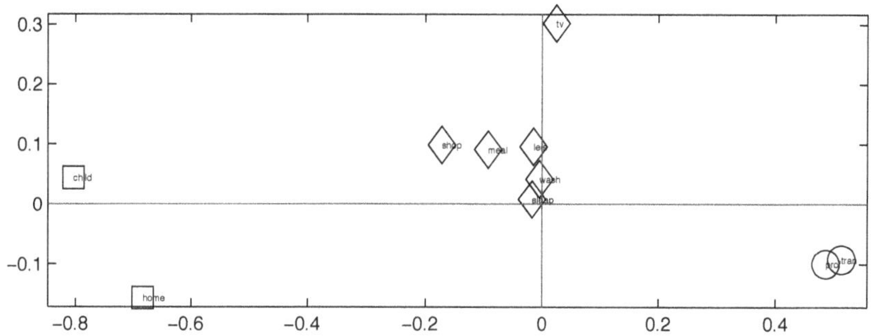


**Figure 1.4.** *Projection of the columns into the factorial plane spawned by the first and second axes*

### 1.10.1. *Choice of model and of the number of classes*

Clustering methods are often justified heuristically, and choosing the "right" method or the "right" number of classes can be a problem that is difficult, and often badly stated. The use of clustering methods based on mixture models allows us to place the problem within the more general framework of the selection of probabilistic models.

In the Bayesian context, choosing the most probable model calls for frequently used selection criteria such as Schwarz's [SCH 78] BIC criterion comprising two terms: the first is likelihood, which tends to favor the more complex model, and the second is a penalizing term, an increasing function of the number of the model's parameters. Worth mentioning is the ICL criterion [BIE 00] which, taking the objective of the clustering into account, generally provides good solutions.

### 1.10.2. *Strategies for use*

Maximizing the likelihood criterion via the EM algorithm or maximizing the clustering likelihood via the CEM algorithm always involves obtaining a series of solutions that see the criterion increase to a local maximum, and which are therefore dependent on the initial position selected by the algorithm. The strategy usually adopted for obtaining a "good" solution is to run the algorithm several times from different starting points and to retain the best solution. For example, see [BIE 03], where some subtle and effective strategies are examined, including an initial phase in which the algorithm is run a large number of times without waiting for complete convergence.

### 1.10.3. *Extension to particular situations*

We have seen that the mixture model in clustering can cope with a variety of situations (spherical or non-spherical classes, equal or unequal proportions, etc.) and deal with both continuous and binary data. In this section, we briefly list some clustering problems that the mixture model approach addresses quite naturally, illustrating its adaptability to particular situations.

*Noisy data:* atypical or outlier data (measurement errors, etc.) generally perturb clustering methods quite considerably. Getting mixture models to take account of noise can be a simple matter, for example by adding a uniformly distributed class or by using distributions less sensitive to atypical elements, such as Laplace distributions.

*Incomplete labeling in discrimination:* in discrimination we often have, in addition to the learning sample whose class is known, a (sometimes large) set of observations whose class is not known. Making use of these unlabeled observations, which can significantly improve the results of the discrimination, can be easily accomplished by introducing observations whose membership to a class is not brought into question during the iterations of the algorithm to the EM and CEM algorithms.

*Spatial data:* the mixture model is based on the hypothesis that the vector $\mathbf{z} = (z_1, \ldots, z_n)$ grouping the classes of the different observations is an independent sample. There are, however, more complex situations, such as the segmentation of pixels in image processing, where this hypothesis must be rejected. In these cases, the mixture model may be extended to the clustering of geographically localized multivariate observations such as hidden Markov fields, so as to include this type of data.

*Block clustering:* the clustering methods described thus far were all designed to classify individuals, or occasionally variables, but there are other methods, often known as block or simultaneous clustering methods, which process the two sets simultaneously and organize the data into homogeneous blocks. Here too, it is possible to extend the use of mixture models [GOV 02] by using a latent block model generalizing the mixture model [GOV 03, GOV 05, GOV 06, GOV 08]. In the following chapters, we will focus on this model.

## 1.11. Conclusion

In this chapter, we have attempted to show the advantages of using mixture models in clustering. This approach provides a general framework capable of taking into account specificities in the data and in the problem. Moreover, a probabilistic model means being able to harness the entire set of statistical results in proposing solutions to difficult problems such as the choice of the model or the number of classes.

Obviously, one of the difficulties with this approach is in deciding whether the selected mixture model is realistic for the data in question. However, as Everitt [EVE 93] has rightly observed, it is not a difficulty specific to this approach. We cannot avoid choosing a method's underlying hypotheses simply by "concealing" them.

Chapter 2

# Model-Based Co-Clustering

The simplest co-clustering approach is to simultaneously perform clustering of rows and columns using a partition z of the set $I$ of rows and a partition w of the set $J$ of columns. In the terminology of Tucker [TUC 64], this approach can be characterized as seeking a partition on both modes of a matrix. As for the partitioning clustering situation, here the most frequent approach is the metric approach that consists of defining a clustering criterion and then finding an algorithm optimizing this criterion. In the first section, we develop a general framework for an approach of this kind proposed by Govaert [GOV 83, GOV 95].

## 2.1. Metric approach

In the context of exploratory analysis, the aim of co-clustering algorithms is to provide an easily interpretable summary, to be able to jointly use factorial methods and clustering methods, to have similar methods adapted to the main data type, and to obtain classical clustering methods when trying to classify one of the two sets. To achieve this aim, the main idea of the proposed methodology is to summarize the initial matrix x by a smaller matrix a defined from a couple of partitions z and w of $I$ and $J$, having the same structure as the initial matrix. For instance, a $1{,}000 \times 200$ binary matrix will be summarized by a $10 \times 5$ binary matrix or a $1{,}000 \times 200$ contingency table by a $10 \times 5$ contingency table. More precisely, the summarization will be made as shown in Figure 2.1 and the function $s$ will depend on the type of data: for binary data, it will be the majority value; for a contingency table, it will be the sum; and for continuous data, it will be the mean.

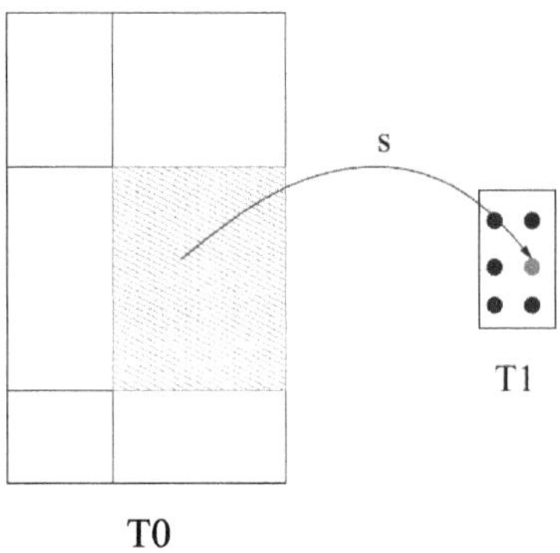


**Figure 2.1.** *Aggregation of data matrix in summary matrix using co-clustering*

In addition to facilitating the interpretation of results, this approach allows us to define the objective function more easily: the objective function $\mathrm{W}(\mathrm{z}, \mathrm{w}, \mathrm{a})$ will be defined using a function $\Delta$ that measures the difference between the two matrices x and a. A more precise definition will depend on the nature of the data.

Several algorithms are able to find a local optimum of the objective function. For instance, Govaert [GOV 83, GOV 95] proposed an iterative algorithm that defines a sequence of partition pairs $(\mathrm{z}, \mathrm{w})$. Starting from an initial partition pair, (z, w), the following procedure is applied: one of these partitions is fixed and a better partition of the other set is searched for. Then, the resulting partition is fixed and a better partition of the first set is searched for. These two steps are repeated until convergence. For finding the better partition at each step, dynamic cluster method [DID 79, CEL 89] can be used. Since this last method is also iterative, the co-clustering algorithms have two levels of iterations. The properties of these types of algorithms are simplicity, speed of convergence and scalability. The drawbacks are due to the fact that they only provide a local optimum. This approach will be discussed in more detail in the following chapters and will lead, respectively, to algorithms CROBIN, CROKI2 and CROEUC for binary, contingency and continuous data.

## 2.2. Probabilistic models

Like the classical clustering, the co-clustering methods described in the previous section are heuristic techniques. To avoid this empirical approach, a number of different probabilistic clustering methods have been proposed, but the use of adapted co-clustering models densities described in this section seems attractive for several reasons: it corresponds to the intuitive idea of a population composed of several blocks, it is strongly linked to classical methods, and it is able to handle a wide variety of special situations in a more or less natural way. These kinds of models are based on a conditional independence assumption that we describe hereafter.

CONDITIONAL INDEPENDENCE.– Given an $n \times d$ data matrix x defined on two sets $I$ and $J$, it is assumed that there exists a partition z on $I$ and a partition w on $J$ such that the univariate random variables $x_{ij}$ are conditionally independent knowing z and w with a parameterized pdf

$f(x_{ij};\alpha_{k\ell})$ if the row $i$ belongs to the cluster $k$ and the column $j$ belong to the cluster $\ell$. Thus, the conditional pdf of x, knowing z and w, can be expressed as

$$\prod_{i,j} f(x_{ij};\alpha_{z_i w_j}) = \prod_{i,j,k,\ell} \{f(x_{ij};\alpha_{k\ell})\}^{z_{ik}w_{j\ell}}. \qquad [2.1]$$

From this hypothesis, three situations can be considered depending on whether $I$ and $J$ are assumed to be random samples or not.

*Block model:* in this situation, the sets $I$ and $J$ are not random samples and statistical units are the elements $x_{ij}$ of the data matrix. Therefore, this model is parameterized by z, w and $\alpha$. These parameters are collectively represented by $\theta$ and the pdf of the data x for this model can be written as

$$f(\mathrm{x};\boldsymbol{\theta}) = \prod_{i,j} f(x_{ij};\alpha_{z_i w_j}) = \prod_{i,j,k,\ell} \{f(x_{ij};\alpha_{k\ell})\}^{z_{ik}w_{j\ell}},$$

where $\boldsymbol{\alpha} = (\alpha_{k\ell}; k = 1,\ldots g, \ell = 1,\ldots,m)$. This model will be studied in Chapter 4 which deals with contingency tables.

*Mixture model:* in this situation, only the set $J$ of rows is regarded as a random sample and the statistical units are rows $\boldsymbol{x}_i$ of the data matrix. Moreover, these statistical units and the row labels, which are latent variables, are assumed to be independent. Therefore, this model is parameterized by $\boldsymbol{\pi} = (\pi_1,\ldots,\pi_g)$ where $\pi_k = P(z_{ik} = 1)$, z, w and $\boldsymbol{\alpha}$ are collectively represented by $\boldsymbol{\theta}$. The pdf of the data x can be written as

$$f(\mathrm{x};\boldsymbol{\theta}) = \prod_i \left( \sum_k \pi_k \prod_{j,\ell} \{f(x_{ij};\alpha_{k\ell})\}^{w_{j\ell}} \right),$$

which is a classical mixture model where, conditional to the component, the variables are independent. This model will be studied in Chapter 5 dealing with continuous variables.

*Latent block model:* in this last situation, the two sets $I$ and $J$ are considered as random samples and the row and column labels become latent variables. The statistical unit is therefore the data matrix, and the remaining chapter will be devoted to this model.

## 2.3. Latent block model

### 2.3.1. *Definition*

To embed the co-clustering in a probabilistic framework, Govaert and Nadif [GOV 03, GOV 05, GOV 06, GOV 07, GOV 08, GOV 10] proposed the latent block model (LBM) and considered different distributions. This model is based on the following assumptions:

– Conditional independence defined in section 2.2.

– Independent latent variables: the labelings $\mathbf{z}_1, \ldots, \mathbf{z}_n, \mathbf{w}_1, \ldots, \mathbf{w}_d$ are considered as latent variables and assumed to be independent

$$p(\mathbf{z}, \mathbf{w}) = p(\mathbf{z})p(\mathbf{w}),\ p(\mathbf{z}) = \prod_i p(z_i) \text{ and } p(\mathbf{w}) = \prod_j p(w_j). \quad [2.2]$$

– For all $i$, the distribution of $p(z_i)$ is the categorical distribution $\mathcal{M}(\pi_1, \ldots, \pi_g)$ and does not depend on $i$. Similarly, for all $j$, the distribution of $p(w_j)$ is the categorical distribution $\mathcal{M}(\rho_1, \ldots, \rho_m)$ and does not depend on $j$.

Therefore, the parameter of the LBM is

$$\boldsymbol{\theta} = (\boldsymbol{\pi}, \boldsymbol{\rho}, \boldsymbol{\alpha}),$$

with $\boldsymbol{\pi} = (\pi_1, \ldots, \pi_g)$ and $\boldsymbol{\rho} = (\rho_1, \ldots, \rho_m)$ where $(\pi_k = P(z_{ik} = 1), k = 1, \ldots, g)$, $(\rho_\ell = P(w_{j\ell} = 1), \ell = 1, \ldots, m)$ are the mixing proportions and $\boldsymbol{\alpha} = (\alpha_{k\ell}; k = 1, \ldots g, \ell = 1, \ldots, m)$ where $\alpha_{k\ell}$ is the parameter of the distribution of block $k, \ell$.

Denoting by $\mathcal{Z}$ and $\mathcal{W}$ the sets of possible labels z for $I$ and w for $J$, the pdf of x can be written as

$$f(\mathbf{x};\boldsymbol{\theta}) = \sum_{(z,w)\in\mathcal{Z}\times\mathcal{W}} p(\mathbf{z},\mathbf{w})f(\mathbf{x}|\mathbf{z},\mathbf{w};\boldsymbol{\theta}) \quad [2.3]$$

$$= \sum_{(z,w)\in\mathcal{Z}\times\mathcal{W}} \prod_i \pi_{z_i} \prod_j \rho_{w_j} \prod_{i,j} f(x_{ij};\alpha_{z_i w_j}) \quad [2.4]$$

$$= \sum_{(z,w)\in\mathcal{Z}\times\mathcal{W}} \prod_{i,k} \pi_k^{z_{ik}} \prod_{j\ell} \rho_\ell^{w_{j\ell}} \prod_{i,j,k,\ell} \{f(x_{ij};\alpha_{k\ell})\}^{z_{ik}w_{j\ell}}. \quad [2.5]$$

The randomized data generation process corresponding to the LBM with parameter $\boldsymbol{\theta} = (\boldsymbol{\pi}, \boldsymbol{\rho}, \boldsymbol{\alpha})$ can be described as follows:

– Generate the labelings $\mathbf{z} = (z_1, \ldots, z_n)$ into $g$ clusters according to the categorical distribution $\boldsymbol{\pi} = (\pi_1, \ldots, \pi_g)$.

– Generate the labelings $\mathbf{w} = (w_1, \ldots, w_d)$ into $m$ clusters according to the categorical distribution $\boldsymbol{\rho} = (\rho_1, \ldots, \rho_m)$.

– Generate for $i = 1, ..., n$ and $j = 1, ..., d$ a real value $x_{ij}$ according to the distribution $f(.;\boldsymbol{\alpha}_{z_i w_j})$.

This model can be represented by a graphical model shown in Figure 2.2.

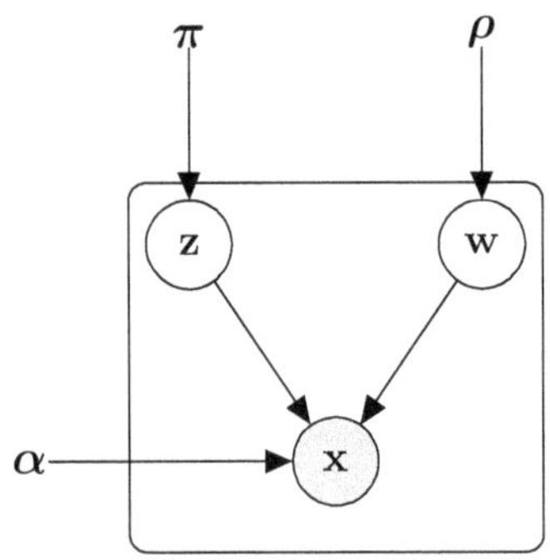


**Figure 2.2.** *Latent block model as a graphical model*

According to the type of data (binary, contingency and continuous) various monodimensional distribution functions can be used. In the following chapters, three distributions will be studied: Bernoulli distributions for binary data, Poisson distributions for contingency data and Gaussian distributions for continuous data, but the model can be extended by taking into account, for example, the outliers by using Student's distributions.

Note that the LBM is dramatically more parsimonious than using a classical mixture model on each of the two sets of rows and columns: for example, with $n = 1{,}000$ rows and $d = 500$ columns and equal class probabilities $\pi_k = 1/g$ and $\rho_\ell = 1/m$, if we need to cluster the data matrix into $g = 4$ clusters of rows and $m = 3$ clusters of columns, the Poisson LBM will involve the estimation of 12 parameters $(\alpha_{k\ell}, k = 1, \ldots, 4, \ell = 1, \ldots, 3)$, instead of $(4 \times 500 + 3 \times 1{,}000) = 5{,}000$ parameters with two mixture models applied on the two sets separately.

Links with other models such as the mixture model, the latent Dirichlet allocation (LDA) model [BLE 03] or the stochastic block model can be established. We study the link with the classical mixture model in the next section.

### 2.3.2. *Link with the mixture model*

It is natural to ask about the relationship between the LBM and the classical mixture model. To this end, it suffices to express the pdf of the data x conditionally on the partition w. This takes the form

$$\begin{aligned} f(\mathbf{x}|\mathbf{w}) &= \sum_{\mathbf{z}} p(\mathbf{z}) f(\mathbf{x}|\mathbf{z},\mathbf{w}) = \sum_{\mathbf{z}} p(\mathbf{z}) \prod_{i,j} f(x_{ij}, \alpha_{z_i w_j}) \\ &= \sum_{\mathbf{z}} \prod_i p(z_i) \prod_{i,j} f(x_{ij}, \alpha_{z_i w_j}) = \sum_{\mathbf{z}} \prod_i \left( p(z_i) \prod_j f(x_{ij}, \alpha_{z_i w_j}) \right) \\ &= \prod_i \sum_k \left( \pi_k \prod_j f(x_{ij}, \alpha_{k w_j}) \right), \end{aligned}$$

which leads to the following properties:

– the component $x_{ij}$ of row $i$ are conditionally independent on $z_{ik}$;

– the rows $\mathrm{x}_i$ of the data matrix x are conditionally independent on w;

– conditional on the partition w, the pdf of the data x is a mixture model;

and, symmetrically:

– the components $x_{ij}$ of column $i$ are conditionally independent on $w_{j\ell}$;

– the columns $\mathrm{x}_j$ of the data matrix x are conditionally independent on z;

– conditional on the partition z, the pdf of the data x is a mixture model.

### 2.3.3. *Log-likelihoods*

In the following, two log-likelihoods will be used: the classical log-likelihood $\mathrm{L}(\boldsymbol{\theta})$, which is the logarithm of the sample distribution $f(\mathrm{x};\boldsymbol{\theta})$ seen as a function of $\boldsymbol{\theta}$, and the *complete-data log-likelihood* function, which is the logarithm of the sample distribution of the latent data, or complete data. For the LBM, the complete data are taken to be the vector $(\mathrm{x}, \mathrm{z}, \mathrm{w})$ where unobservable vectors z and w are the labels. The complete data log-likelihood can therefore be written as

$$\begin{aligned}
\mathrm{L_C}(\mathbf{z}, \mathrm{w}, \boldsymbol{\theta}) &= \mathrm{L}(\boldsymbol{\theta}; \mathrm{x}, \mathrm{z}, \mathrm{w}) = \log f(\mathrm{x}, \mathrm{z}, \mathrm{w}; \boldsymbol{\theta}) \\
&= \log\{p(\mathbf{z}; \boldsymbol{\theta}) p(\mathrm{w}; \boldsymbol{\theta}) f(\mathrm{x}|\mathbf{z}, \mathrm{w}; \boldsymbol{\alpha})\} \\
&= \log p(\mathbf{z}; \boldsymbol{\theta}) + \log p(\mathrm{w}; \boldsymbol{\theta}) + \log f(\mathrm{x}|\mathbf{z}, \mathrm{w}; \boldsymbol{\alpha}) \\
&= \log \prod_{i,k} \pi_k^{z_{ik}} + \log \prod_{j,\ell} \rho_\ell^{w_{j\ell}} + \log f(\mathrm{x}|\mathbf{z}, \mathrm{w}; \boldsymbol{\alpha}),
\end{aligned}$$

and, finally, the complete-data log-likelihood becomes

$$L_C(\mathbf{z}, \mathbf{w}, \boldsymbol{\theta}) = \sum_{i,k} z_{ik} \log \pi_k + \sum_{j,\ell} w_{j\ell} \log \rho_\ell + \sum_{i,j,k,\ell} z_{ik} w_{j\ell} \log f(x_{ij}; \alpha_{k\ell}). \quad [2.6]$$

Note that this complete-data log-likelihood divides into three terms: the first term depends on proportions of row clusters, the second term depends on proportions of column clusters and the third term depends on the pdf of each block or bicluster.

### 2.3.4. *A complex model*

The LBM involves a double missing data structure, namely z and w, which makes statistical inference more difficult than for the standard mixture model. In this section, we list some of these difficulties.

#### 2.3.4.1. *Computing the likelihood*

Even for small tables, computing this likelihood (or its logarithm) is difficult. For instance, with a data matrix $20 \times 20$ with $g = 2$ and $m = 2$, it requires the calculation of $gn \times md = 1{,}600$ terms. As a consequence, even though the parameter $\boldsymbol{\theta}$ is properly estimated, computing penalized model selection criteria such as the Akaike information criterion (AIC) or Bayesian information criterion (BIC) is challenging. Moreover, these difficulties of computing relevant model selection criteria are increased by the fact that the LBM statistical units could be defined in several different ways.

#### 2.3.4.2. *Classification step*

With the parameter $\boldsymbol{\theta}$ being fixed, the problem is to determine the "best" partitions z and w, i.e. the pair of partitions $\mathbf{z}, \mathbf{w}$ maximizing the posterior probability $f(\mathbf{z}, \mathbf{w}|\mathbf{x}, \boldsymbol{\theta})$ (MAP). Knowing that

$$f(\mathbf{z}, \mathbf{w}|\mathbf{x}, \boldsymbol{\theta}) = \frac{f(\mathbf{x}, \mathbf{z}, \mathbf{w}; \boldsymbol{\theta})}{f(\mathbf{x}; \boldsymbol{\theta})},$$

the couple $(\mathbf{z}, \mathbf{w})$ is obtained by maximizing the complete-data log-likelihood $\mathrm{L_C}(\boldsymbol{\theta}, \mathbf{z}, \mathbf{w}) = \log f(\mathbf{x}, \mathbf{z}, \mathbf{w}; \boldsymbol{\theta})$ in $\mathbf{z}$ and $\mathbf{w}$. This result generalizes the case of the simple mixing model where the MAP is obtained by maximizing $\mathrm{L_C}(\boldsymbol{\theta}, \mathbf{z})$ in $\mathbf{z}$, i.e. $\log \prod_i f(\boldsymbol{x}_i, z_i)$, which leads to placing each $\boldsymbol{x}_i$ in the class $z_i$ that maximizes

$$f(\boldsymbol{x}_i, z_i) \propto \frac{f(\boldsymbol{x}_i, z_i)}{f(\boldsymbol{x}_i)} = f(z_i|\boldsymbol{x}_i).$$

Unfortunately, in the latent block situation, the maximization of the complete-data log-likelihood $\mathrm{L_C}(\mathbf{z}, \mathbf{w}, \boldsymbol{\theta})$ does not lead to explicit formulas and iterative algorithms are needed.

2.3.4.3. *Error rate*

One characteristic of a mixture model is the degree of mixing among the components. In the classical situation, this concept of cluster separation can be visualized, for instance, by using principal component analysis, but this concept of cluster separation is difficult to apply to the LBMs. Another solution is to compute the true error rate associated with the model, which is defined as the expectation of the misclassification probability $\mathbb{E}(\delta((\mathbf{z}, \mathbf{w}), d(\mathbf{x}))$ where $\mathbf{z}$, $\mathbf{w}$ and $\mathbf{x}$ are the random variables associated with the LBM, $d$ is the optimal Bayes' rule $d(\mathbf{x}) = (\mathbf{z}', \mathbf{w}') = \arg\max_{\mathbf{z}, \mathbf{w}} p(\mathbf{z}, \mathbf{w}|\mathbf{x})$ associated with this model and $\delta$ is the error rate.

This expectation is generally difficult to compute theoretically, and Monte Carlo simulations are used to estimate it by the proportion of misclassification, for instance in the classical clustering situation, between the partition simulated with those we obtained by applying a classification step. This proportion of misclassification can be defined as follows: if $C$ is the confusion matrix between the two partitions, relabel the components of the partition $\mathbf{z}'$ such that the trace of matrix $C$ is maximal, then compute $e(\mathbf{z}, \mathbf{z}') = 1 - \frac{1}{n}\sum_{i,k} z_{ik} z'_{ik}$. This definition can be extended to the comparison of two pairs of partitions $\mathbf{u} = (\mathbf{z}, \mathbf{w})$ and $\mathbf{u}' = (\mathbf{z}', \mathbf{w}')$ as follows

$$\delta(\mathbf{u}, \mathbf{u}') = \delta((\mathbf{z}, \mathbf{w}), (\mathbf{z}', \mathbf{w}')) = 1 - \frac{1}{nd} \sum_{i,j,k,\ell} u_{ijk\ell} u'_{ijk\ell},$$

where $u_{ijk\ell} = z_{ik} w_{j\ell}$ and $u'_{ijk\ell} = z'_{ik} w'_{j\ell}$, and it can be shown that

$$\delta((\mathbf{z}, \mathbf{w}), (\mathbf{z}', \mathbf{w}')) = e(\mathbf{z}, \mathbf{z}') + e(\mathbf{w}, \mathbf{w}') - e(\mathbf{z}, \mathbf{z}') \times e(\mathbf{w}, \mathbf{w}').$$

However, this approach is challenging in the LBM situation:

– The classification step is not direct as in a classical mixture model situation, and an iterative algorithm must be used (see section 2.3.4.2).

– The error rate depends not only on the parameter $\theta$ but also on the sizes $n$ and $d$ of the data array, which makes Monte Carlo estimation of the error rate impractical. This unusual phenomenon can be seen in Figure 2.3: for a given distribution, the error rates decrease as the table size increases. The error decrease reflects a property of the table distribution: its size plays an important role in setting the Bayes' risk, i.e. the intrinsic difficulty of the task. To understand this phenomenon, consider the representation of an $n \times d$ table as $n$ $d$-dimensional vectors. The overall dissimilarity among vectors will grow as $d$ grows. Figure 2.4

shows the principal components of such vectors, extracted from two tables with $d = 50$ in the left plot and $d = 500$ in the right plot. The distribution describing the classes and their probabilities is identical in both plots, but while the clusters in dimension $d = 50$ highly overlap, they are well separated in $d = 500$. Figures 2.3 and 2.4 illustrate a systematic but little-known property of block clustering: the distribution of the table entries being fixed, the Bayes' risk decreases with the table size. More formal arguments, already developed for the more constrained stochastic block model [CEL 12], could be transposed onto co-clustering. Intuitively, this decrease can be understood by considering that the table enlargement in one dimension results in more redundancy in the other dimension.

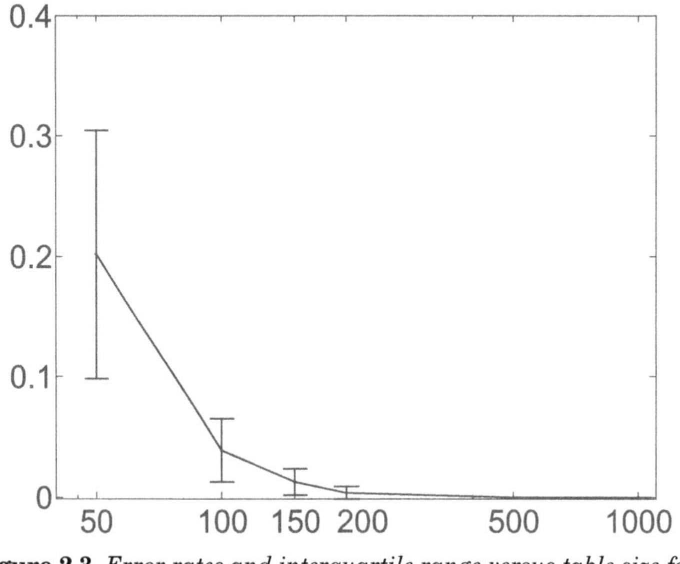


**Figure 2.3.** *Error rates and interquartile range versus table size for square data tables ($m = n$) with $3 \times 3$ clusters whose entries are generated from the same distribution*

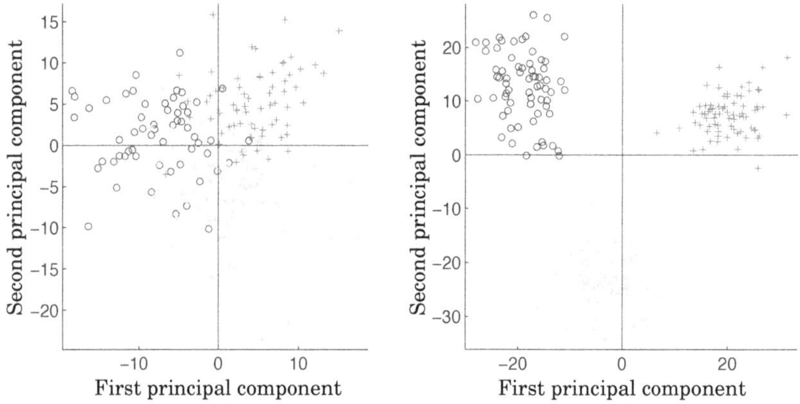


$n = 200$, $d = 50$ $\qquad$ $n = 200$, $d = 500$

**Figure 2.4.** *Projections of the rows of two data tables on the first two column eigenvectors. The distribution of table entries is identical, but table sizes differ, with* $n = 200$ *rows and* $d = 50$ *(left) or* $d = 500$ *(right)*

## 2.4. Maximum likelihood estimation and algorithms

As for the classical mixture model, the maximization of the likelihood is not straightforward and the expectation–maximization (EM) algorithm [DEM 77], which maximizes the log-likelihood $\mathrm{L}(\boldsymbol{\theta})$ iteratively by maximizing the conditional expectation of the *complete-data log-likelihood* given a previous current estimate $\boldsymbol{\theta}^{(c)}$ and the data x, can be considered. Unfortunately, difficulties arise owing to the dependence structure in the model. The LBM involves a double missing data structure, namely z and w, which makes statistical inference more difficult than for the standard mixture model. As previously seen, computing the likelihood [2.3] or its logarithm is difficult. In the same manner, deriving the maximum likelihood estimator with the EM algorithm is challenging. As a matter of fact, the E step requires the computation of the joint conditional distributions of the missing labels $p(z_{ik}w_{j\ell} = 1|\mathrm{x}; \boldsymbol{\theta})$ for $i \in I$, $j \in J$, $k \in K$ and $\ell \in L$, $\boldsymbol{\theta}$ being a current value of the parameter. Thus, the E step, which computes the conditional expectation of the complete data log-likelihood given a previous current estimate $\boldsymbol{\theta}^{(c)}$ and x

$$Q(\boldsymbol{\theta},\boldsymbol{\theta}^{(c)}) = \sum_{i,k} p(z_{ik}=1|\mathbf{x},\boldsymbol{\theta}^{(c)}) \log \pi_k + \sum_{j,\ell} p(w_{j\ell}=1|\mathbf{x},\boldsymbol{\theta}^{(c)}) \log \rho_\ell$$
$$+ \sum_{i,j,k,\ell} p(z_{ik}w_{j\ell}=1|\mathbf{x},\boldsymbol{\theta}^{(c)}) \log f(x_{ij};\alpha_{k\ell}),$$

involves computing too many terms that cannot be factorized, such as for a standard mixture, due to the conditional dependence on the observations of the row and column labels.

To solve this problem, an approximation using the Neal and Hinton interpretation [NEA 98] of the EM algorithm can be proposed. In Chapter 1, it was seen that, for mixture models, the EM algorithm can be viewed as an alternating algorithm. To extend this property to all models that can be dealt with by using the EM algorithm, Neal and Hinton propose to define the following function

$$\mathrm{F_C}(R;\boldsymbol{\theta}) = \mathbb{E}_R(\mathrm{L_C}(\mathbf{y};\boldsymbol{\theta})|\mathbf{x}) + \mathrm{H}(R), \qquad [2.7]$$

where y is the complete data, $R$ is a probability over the space of complete data and $\mathrm{H}(R)$ is the entropy of the distribution $R$. We can also relate $\mathrm{F_C}$ to the Kullback–Liebler divergence between $R$ and $P_{\boldsymbol{\theta}}$ defined by $P_{\boldsymbol{\theta}}(\mathbf{y}) = p(\mathbf{y}|\mathbf{x},\boldsymbol{\theta})$ as follows

$$\mathrm{F_C}(R,\boldsymbol{\theta}) = \mathrm{L}(\theta) - \mathrm{KL}(R,P_{\boldsymbol{\theta}}). \qquad [2.8]$$

The alternated optimization of the $\mathrm{F_C}$ function is therefore simple to set up: for fixed $\boldsymbol{\theta}$, the maximization of equation [2.8] yields to the minimization of $\mathrm{KL}(R,\mathbb{P}_\theta)$ and therefore to $R = P_{\boldsymbol{\theta}}$; for fixed $R$, the maximization of equation [2.7] shows that $\boldsymbol{\theta}$ must maximize the expectation $E_R(\mathrm{L_C}(\mathbf{y},\boldsymbol{\theta})|\mathbf{x})$. These two steps are precisely those of the EM algorithm. Moreover, after the first step, we have $\mathrm{F_C}(R,\boldsymbol{\theta}) = \mathrm{F_C}(P_{\boldsymbol{\theta}},\boldsymbol{\theta}) = L(\boldsymbol{\theta})$, which shows that each iteration increases the log-likelihood. The relation between the log-likelihood and the fuzzy criterion $\mathrm{F_C}$ can be written as $\mathrm{L}(\boldsymbol{\theta}) = \arg\max_R \mathrm{F_C}(R, \boldsymbol{\theta})$.

For the LB model, $R$ is a distribution on $\mathcal{Z} \times \mathcal{W}$ and the first step leads to taking $R = P(\mathbf{z}, \mathrm{w}|\mathrm{x}, \boldsymbol{\theta})$, which is, obviously, always intractable but, to derive the maximum likelihood estimate of $\boldsymbol{\theta}$, two approximations can be proposed: a variational EM approach and a classification EM approach.

### 2.4.1. *Variational EM approach*

A variational approximation of the EM algorithm, denoted as VEM in this book, can be proposed by imposing that the distribution $R$ of the labels is assumed to be independent

$$R(\mathbf{z}, \mathrm{w}) = R(\mathbf{z})R(\mathrm{w}), \quad R(\mathbf{z}) = \prod_i q(z_i)$$

and

$$R(\mathrm{w}) = \prod_j q(w_j). \qquad [2.9]$$

Denoting $\widetilde{z}_{ik} = R(z_{ik} = 1)$, $\widetilde{\mathrm{w}}_{j\ell} = R(w_{j\ell} = 1)$, $\widetilde{z}_{.k} = \sum_i \widetilde{z}_{ik}$ and $\widetilde{\mathrm{w}}_{.\ell} = \sum_j \widetilde{\mathrm{w}}_{j\ell}$, the expectation of complete-data log-likelihood and the entropy become

$$\mathbb{E}_R(\mathrm{L_C}(\mathbf{y}; \boldsymbol{\theta})|\mathrm{x}) = \sum_k \widetilde{z}_{.k} \log \pi_k + \sum_\ell \widetilde{\mathrm{w}}_{.\ell} \log \rho_\ell + \sum_{i,j,k,\ell} \widetilde{z}_{ik}\widetilde{\mathrm{w}}_{j\ell} \log f(x_{ij}, \boldsymbol{\alpha}_{k\ell})$$

and

$$\mathrm{H}(P) = \mathrm{H}(\widetilde{\mathbf{z}}) + \mathrm{H}(\widetilde{\mathbf{w}}),$$

where $\mathrm{H}(\widetilde{\mathbf{z}}) = -\sum_{i,k} \widetilde{z}_{ik} \log \widetilde{z}_{ik}$ and $H(\widetilde{\mathbf{w}}) = -\sum_{j,\ell} \widetilde{\mathrm{w}}_{j\ell} \log \widetilde{\mathrm{w}}_{j\ell}$. Therefore, the fuzzy clustering criterion [2.7] can be written as

$$\mathrm{F_C}(\widetilde{\mathbf{z}}, \widetilde{\mathbf{w}}; \boldsymbol{\theta}) = \mathrm{L_C}(\widetilde{\mathbf{z}}, \widetilde{\mathbf{w}}, \boldsymbol{\theta}) + \mathrm{H}(\widetilde{\mathbf{z}}) + \mathrm{H}(\widetilde{\mathbf{w}}), \qquad [2.10]$$

where $\mathrm{L_C}$ is the fuzzy complete-data log-likelihood associated with the block latent model depending on three terms as in equation [2.6] but weighted by posterior probabilities

$$\mathrm{L_C}(\widetilde{\mathbf{z}}, \widetilde{\mathbf{w}}; \boldsymbol{\theta}) = \sum_k \widetilde{z}_{.k} \log \pi_k + \sum_\ell \widetilde{\mathrm{w}}_{.\ell} \log \rho_\ell + \sum_{i,j,k,\ell} \widetilde{z}_{ik} \widetilde{\mathrm{w}}_{j\ell} \log f(x_{ij}, \boldsymbol{\alpha}_{k\ell}).$$

The objective of the variational approach is therefore to maximize the function $\mathrm{F_C}$. In doing so, we have replaced the maximization of the likelihood by the maximization of an approximation of this likelihood defined by

$$\widetilde{\mathrm{L}}(\boldsymbol{\theta}) = \arg\max_{\widetilde{\mathbf{z}}, \widetilde{\mathbf{w}}} \mathrm{F_C}(\widetilde{\mathbf{z}}, \widetilde{\mathbf{w}}, \boldsymbol{\theta}).$$

The maximization of the function $\mathrm{F_C}(\widetilde{\mathbf{z}}, \widetilde{\mathrm{w}}, \boldsymbol{\delta})$ can be obtained by alternating the following three computations:

1) $\widetilde{\mathbf{z}} = \arg\max_{\widetilde{\mathbf{z}}} F_C(\widetilde{\mathbf{z}}, \widetilde{\mathrm{w}}, \boldsymbol{\theta})$: equations [2.10] and [2.11] lead, for all $i$, to

$$\widetilde{\mathbf{z}} = \arg\max_{\widetilde{\mathbf{z}}} \sum_k \widetilde{z}_{ik} \left( \log \pi_k + \sum_{j,\ell} \widetilde{\mathrm{w}}_{j\ell} \log f(x_{ij}, \boldsymbol{\alpha}_{k\ell}) \right)$$

under the constraint $\sum_k \widetilde{z}_{ik} = 1$ and therefore to

$$\widetilde{z}_{ik} \propto \pi_k \exp(\sum_{j,\ell} \widetilde{\mathrm{w}}_{j\ell} \log f(x_{ij}, \boldsymbol{\alpha}_{k\ell})).$$

2) $\widetilde{\mathrm{w}} = \arg\max_{\widetilde{\mathbf{z}}} F_C(\widetilde{\mathbf{z}}, \widetilde{\mathrm{w}}, \boldsymbol{\theta})$: similarly, we obtain

$$\widetilde{\mathrm{w}}_{j\ell} \propto \rho_\ell \exp(\sum_{i,j} \widetilde{z}_{ik} \log f(x_{ij}, \boldsymbol{\alpha}_{k\ell})).$$

3) $\boldsymbol{\theta} = \arg\max_{\boldsymbol{\theta}} F_C(\widetilde{\mathbf{z}}, \widetilde{\mathrm{w}}, \boldsymbol{\theta})$: equations [2.10] and [2.11] lead to

$$\boldsymbol{\pi} = \arg\max_{\boldsymbol{\pi}} \sum_k \widetilde{z}_k \log \pi_k \quad \text{and} \quad \boldsymbol{\rho} = \arg\max_{\boldsymbol{\rho}} \sum_\ell \widetilde{\mathrm{w}}_\ell \log \rho_\ell$$

and therefore to

$$\pi_k = \frac{\widetilde{z}_{.k}}{n} \quad \forall k \qquad \text{and} \qquad \rho_\ell = \frac{\widetilde{\mathrm{w}}_{.\ell}}{d} \quad \forall \ell.$$

For the block parameters, the computations of the $\alpha_{k\ell}$'s, defined by $\arg\max_{\alpha_{k\ell}} \sum_{i,j} \widetilde{z}_{ik} \widetilde{\mathrm{w}}_{j\ell} \log f(x_{ij}, \boldsymbol{\alpha}_{k\ell})\ \forall k, \ell$, will depend on the distribution of the component and will be studied in the following chapters.

Different strategies of alternated optimization can be proposed. One of them, denoted as LBVEM algorithm and described in algorithm [2.1], has been proved to give relevant estimates of the LBM model in different contexts; see, for instance, [GOV 08] for continuous or binary data.

After we fit the LBM to estimate $\boldsymbol{\theta}$, we can give an outright or hard clustering of these data by assigning each observation to the component of the mixture which it has the highest posterior probability of belonging to, that is to say to compute the maximum *a posteriori* (MAP). Unfortunately, this maximization does not lead to explicit formulas and iterative algorithms are needed. Here, a simple solution is to assign each row or column to the component which maximizes the probabilities $\widetilde{z}_{ik}$ or $\widetilde{\mathrm{w}}_{j\ell}$ obtained at the end of the VEM algorithm.

**Algorithm 2.1** LBVEM

**input:** x, $g$, $m$
**initialization:** $\widetilde{\mathrm{z}}$, $\widetilde{\mathrm{w}}$, $\pi_k = \frac{\widetilde{z}_{.k}}{n}$, $\rho_\ell = \frac{\widetilde{\mathrm{w}}_{.\ell}}{d}$,
$\alpha_{k\ell} = \arg\max_{\alpha_{k\ell}} \sum_{i,j} \widetilde{z}_{ik}\widetilde{\mathrm{w}}_{j\ell} \log f(x_{ij}, \boldsymbol{\alpha}_{k\ell})$
**repeat**
  **repeat**
    **step 1.** $\widetilde{z}_{ik} \propto \pi_k \exp(\sum_{j,\ell} \widetilde{\mathrm{w}}_{j\ell} \log f(x_{ij}, \boldsymbol{\alpha}_{k\ell}))$
    **step 2.** $\pi_k = \frac{\widetilde{z}_{.k}}{n}$,
    $\alpha_{k\ell} = \arg\max_{\alpha_{k\ell}} \sum_{i,j} \widetilde{z}_{ik}\widetilde{\mathrm{w}}_{j\ell} \log f(x_{ij}, \boldsymbol{\alpha}_{k\ell})$
  **until** convergence
  **repeat**
    **step 3.** $\widetilde{\mathrm{w}}_{j\ell} \propto \rho_\ell \exp(\sum_{i,k} \widetilde{z}_{ik} \log f(x_{ij}, \boldsymbol{\alpha}_{k\ell}))$
    **step 4.** $\rho_\ell = \frac{\widetilde{\mathrm{w}}_{.\ell}}{d}$,
    $\alpha_{k\ell} = \arg\max_{\alpha_{k\ell}} \sum_{i,j} \widetilde{z}_{ik}\widetilde{\mathrm{w}}_{j\ell} \log f(x_{ij}, \boldsymbol{\alpha}_{k\ell})$
  **until** convergence
**until** convergence
**return** $\boldsymbol{\pi}$, $\boldsymbol{\rho}$, $\boldsymbol{\alpha}$

### 2.4.2. *Classification EM approach*

Another approximation of the EM algorithm can be obtained by replacing, in equation [2.7], the constraint [2.9] with the constraint

$$p(z_{ik} = 1, w_{j\ell} = 1) \in \{0, 1\}.$$

In doing so, the fuzzy partitions $\widetilde{\mathrm{z}}$ and $\widetilde{\mathrm{w}}$ are replaced by the "hard" partitions z and w and, as the entropies H(z) and H(w) become zero, the objective of this approach is therefore to maximize the complete-data log-likelihood, also called classification log-likelihood,

$$\mathrm{L_C}(\mathbf{z}, \mathrm{w}; \boldsymbol{\theta}) = \sum_{i,k} z_{ik} \log \pi_k + \sum_{j,\ell} w_{j\ell} \log \rho_\ell + \sum_{i,j,k,\ell} z_{ik} w_{j\ell} f(x_{ij}, \alpha_{k\ell}). \qquad [2.11]$$

This optimization can be performed by using the classification EM (CEM) algorithm [CEL 92]. It consists of inserting a classification step between E and M steps. Note that equation [2.11] is just a hard version of the fuzzy version expressed in equation [2.11]. The principal steps of the algorithm, that we refer to as LBCEM, are reported in algorithm 2.2.

**Algorithm 2.2** LBCEM

**input:** x, $g$, $m$
**initialization:** z, w, $\pi_k = \frac{z_{.k}}{n}$, $\rho_\ell = \frac{w_{.\ell}}{d}$,
$\alpha_{k\ell} = \arg\max_{\alpha_{k\ell}} \sum_{i,j} z_{ik} w_{j\ell} \log f(x_{ij}, \boldsymbol{\alpha}_{k\ell})$
**repeat**
  **repeat**
    **step 1.** $z_i = \arg\max_k \left(\sum_{j,\ell} w_{j\ell} \log f(x_{ij}, \boldsymbol{\alpha}_{k\ell}) + \log \pi_k\right)$
    **step 2.** $\pi_k = \frac{z_{.k}}{n}$,
    $\alpha_{k\ell} = \arg\max_{\alpha_{k\ell}} \sum_{i,j} z_{ik} w_{j\ell} \log f(x_{ij}, \boldsymbol{\alpha}_{k\ell})$
  **until** convergence
  **repeat**
    **step 3.** $w_j = \arg\max_\ell \left(\sum_{i,k} z_{ik} \log f(x_{ij}, \boldsymbol{\alpha}_{k\ell}) + \log \rho_\ell\right)$
    **step 4.** $\rho_\ell = \frac{\widetilde{w}_{.\ell}}{d}$,
    $\alpha_{k\ell} = \arg\max_{\alpha_{k\ell}} \sum_{i,j} z_{ik} w_{j\ell} \log f(x_{ij}, \boldsymbol{\alpha}_{k\ell})$
  **until** convergence
**until** convergence
**return** $\pi$, $\rho$, $\alpha$, z, w

### 2.4.3. *Stochastic EM-Gibbs approach*

An alternative will be to replace the EM algorithm with the Gibbs sampling algorithm that, unlike the EM algorithm, may be implemented accurately without problems. Moreover, this approach, which does not stop at the first encountered fixed point of EM [MCL 07], is a possible way to attenuate the dependence of VEM on its initial values. The difficulty is that it is a stochastic algorithm that requires a large number of iterations and does not directly provide a pointwise estimate.

The basic idea of these stochastic EM algorithms is to incorporate a stochastic step between the E and M steps where the missing data are simulated according to their conditional distribution, knowing the observed data and a current estimate of the model parameters. For the LBM, it is not possible to simulate in a single exercise the missing labels z and w, and a Gibbs sampling scheme is required to simulate the couple $(\mathbf{z}, \mathbf{w})$. The SEM-Gibbs algorithm (algorithm 2.3) is a simple adaptation to the LBVEM of the standard SEM algorithm of [CEL 85].

**Algorithm 2.3** LBSEM

**input:** $\mathbf{x}$, $g$, $m$
**initialization:** $\mathbf{z}$, $\mathbf{w}$, $\pi_k = \frac{z_{.k}}{n}$, $\rho_\ell = \frac{w_{.\ell}}{d}$,
$\alpha_{k\ell} = \arg\max_{\alpha_{k\ell}} \sum_{i,j} z_{ik} w_{j\ell} \log f(x_{ij}, \boldsymbol{\alpha}_{k\ell})$
**repeat**
  **repeat**
    **step 1.** $\widetilde{z}_{ik} \propto \pi_k \exp(\sum_{j,\ell} w_{j\ell} \log f(x_{ij}, \boldsymbol{\alpha}_{k\ell}))$
    **step 1'.** simulation of $z_i$ according to $\mathcal{M}(\widetilde{z}_{i1}, \ldots, \widetilde{z}_{ig})$
    **step 2.** $\pi_k = \frac{z_{.k}}{n}$,
    $\alpha_{k\ell} = \arg\max_{\alpha_{k\ell}} \sum_{i,j} z_{ik} w_{j\ell} \log f(x_{ij}, \boldsymbol{\alpha}_{k\ell})$
  **until** convergence
  **repeat**
    **step 3.** $\widetilde{\mathrm{w}}_{j\ell} \propto \rho_\ell \exp(\sum_{i,k} z_{ik} \log f(x_{ij}, \boldsymbol{\alpha}_{k\ell}))$
    **step 3'.** simulation of $w_j$ according to $\mathcal{M}(\widetilde{\mathrm{w}}_{j1}, \ldots, \widetilde{\mathrm{w}}_{jm})$
    **step 4.** $\rho_\ell = \frac{w_{.\ell}}{d}$,
    $\alpha_{k\ell} = \arg\max_{\alpha_{k\ell}} \sum_{i,j} z_{ik} w_{j\ell} \log f(x_{ij}, \boldsymbol{\alpha}_{k\ell})$
  **until** convergence
**until** convergence
**return** $\pi$, $\rho$, $\alpha$

Note that the formulas of VEM and SEM-Gibbs are essentially the same, except that the probabilities $\widetilde{z}_{ik}$ and $\widetilde{\mathrm{w}}_{j\ell}$

are replaced with binary indicator values $z_{ik}$ and $w_{j\ell}$. However, it is not the only difference between VEM and SEM-Gibbs: while VEM is based on the variational approximation of the LBM, SEM-Gibbs uses no approximation, but runs a Gibbs sampler to simulate the unknown labels with their conditional distribution knowing the observations and a current estimation of the parameters. SEM-Gibbs does not increase the log-likelihood at each iteration. It generates an irreducible Markov chain with a unique stationary distribution that is expected to be concentrated about the maximum likelihood parameter estimate. Thus, a natural estimate of $\boldsymbol{\theta}$ derived from SEM-Gibbs is the mean $\bar{\boldsymbol{\theta}}$ of $(\boldsymbol{\theta}^{(c)}; c = B+1, \ldots, B+C)$ obtained after a burn-in period of length $B$. Numerical experiments presented in [KER 13] for binary data show that SEM-Gibbs is by far less sensitive to starting values than VEM. Those results have led them to advocate initializing the VEM algorithm with the SEM-Gibbs mean parameter estimate $\bar{\boldsymbol{\theta}}$ to obtain a good approximation of the maximum likelihood estimate for the LBM.

## 2.5. Bayesian approach

Bayesian inference in statistics can be regarded as a well-grounded tool for regularizing the maximum likelihood estimates in a poorly posed setting. In the LBM setting, Bayesian inference could be considered as useful in avoiding spurious solutions and thus attenuate the "empty cluster" problem. Using Bayesian inference from a regularization perspective, the model parameter may be estimated by maximizing the posterior probability $p(\boldsymbol{\theta}|\mathrm{x})$; it leads to the so-called MAP estimate

$$\widehat{\boldsymbol{\theta}}_{\mathrm{MAP}} = \arg\max_{\boldsymbol{\theta}} p(\boldsymbol{\theta}|\mathbf{x}).$$

The Bayes' formula

$$\log p(\boldsymbol{\theta}|\mathbf{x}) = \log p(\mathbf{x}|\boldsymbol{\theta}) + \log p(\boldsymbol{\theta}) - \log p(\mathbf{x})$$

allows us to straightforwardly define an EM algorithm for the computation of the MAP estimate

– The E step relies on the computation of the conditional expectation of the complete log-likelihood $R(\boldsymbol{\theta}, \boldsymbol{\theta}^{(c)})$ as for the maximum likelihood estimator

$$R(\boldsymbol{\theta}, \boldsymbol{\theta}^{(c)}) = \mathbb{E}\big(\log p(\mathbf{x}, \mathbf{z}, \mathbf{w}|\boldsymbol{\theta})|\mathbf{x}, \boldsymbol{\theta}^{(c)}\big).$$

– The M step differs in that the objective function for the maximization process is equal to the $R(\boldsymbol{\theta}, \boldsymbol{\theta}^{(c)})$ function augmented by the logarithm of the prior probability

$$\boldsymbol{\theta}^{(c+1)} = \arg\max_{\boldsymbol{\theta}} \big(R(\boldsymbol{\theta}, \boldsymbol{\theta}^{(c)}) + \log p(\boldsymbol{\theta})\big).$$

This M Step forces an increase in the log-posterior function $p(\boldsymbol{\theta}|\mathbf{x})$ [MCL 07, Chapter 6, p. 231]. For LBM, the Bayesian approach combined with the variational approximation leads to the V-Bayes algorithm. This approach will be discussed, in more detail for binary and categorical data in Chapter 3.

## 2.6. Conclusion and miscellaneous developments

In this chapter, we developed the LBM and three algorithms enabling us to estimate the parameters of this model and leading to co-clustering. Note that LBVEM and LBCEM are two iterative algorithms whose convergence in terms of $\mathrm{F_C}(\widetilde{\mathbf{z}}, \widetilde{\mathrm{w}}; \boldsymbol{\theta})$ or $\mathrm{L_C}(\mathbf{z}, \mathrm{w}; \boldsymbol{\theta})$ is guaranteed. They are simple to implement and scalable. Parsimony of the model and its flexibility in adapting to different types of data, as we will see in the remainder of book, are particularly beneficial in the context of data mining. These algorithms also easily avoid the sparsity problem because they work on intermediate matrices that are compressed. We will discuss this point in detail in Chapter 3. The data are called sparse when the proportion of zeros of the matrix is very important. For storing and manipulating such matrices, it is beneficial and often necessary to use specialized algorithms and data structures that take advantage of the sparse structure of the matrix. The sparse data structure represents a matrix in space proportional to the number of non-zero entries, and most of the operations compute sparse results in time proportional to the number of arithmetic operations on non-zeros (see, for instance, [GIL 92]). In this situation, the algorithms described in this chapter are able to take into account this type of data in a simple and effective way.

However, there are still some challenges to be overcome with this model, such as the choice of the number of clusters $(g,m)$. Choosing relevant numbers of clusters in an LBM is obviously very important. However, this model selection problem is more difficult than in a classical mixture model for several reasons. First, there is a couple $(g, m)$ of number of clusters to be selected. Second, penalized likelihood criteria such as $\mathrm{AIC}$ or $\mathrm{BIC}$ are not directly available since computing the maximized likelihood is not feasible. Third, determining the number of statistical units of an LBM could be questionable. In the next chapter, it will be shown how this problem can be solved.

Before considering the estimation problem, it is important to analyze the model generic identifiability [FRÜ 06, pp. 21–23]. Obviously, LBM, as a mixture model, is not identifiable due to invariance to relabeling the blocks, but it is of no importance when used for the maximum likelihood estimation. The identifiability will depend on the type of data and will be studied in more detail for the binary and categorical situations in the next chapter.

# Chapter 3

# Co-Clustering of Binary and Categorical Data

To justify and detail the search for simultaneous partitions of binary data, let us cite some ideas proposed by Lerman about the notion of polythetic cluster [LER 81]. He first recalled the notion of polythetic cluster: "A polythetic cluster $G$ of a natural clustering refers to a subset $B$ of attributes in such a way that: each element of the cluster has an important proportion of attributes from $B$; each attribute of $B$ is present in an important proportion; an attribute is not necessarily shared by all elements of $G$". Lerman generalized this notion: "In the more general situation of a good clustering on $E$ with a good clustering on $A$, each cluster $E_k$ of the partition $(E_1, ..., E_g)$ corresponds to the union $B$ of clusters of the partition $(A_1, ..., A_m)$. Conversely, each cluster $A_\ell$ corresponds to the union $G$ of clusters of the partition $(E_1, ..., E_g)$". This situation may be represented by a binary table reorganized according to row and column partitions (Figure 3.1) where the shaded blocks correspond to regions with high density of ones and the unshaded blocks correspond to regions with high density of zeros.

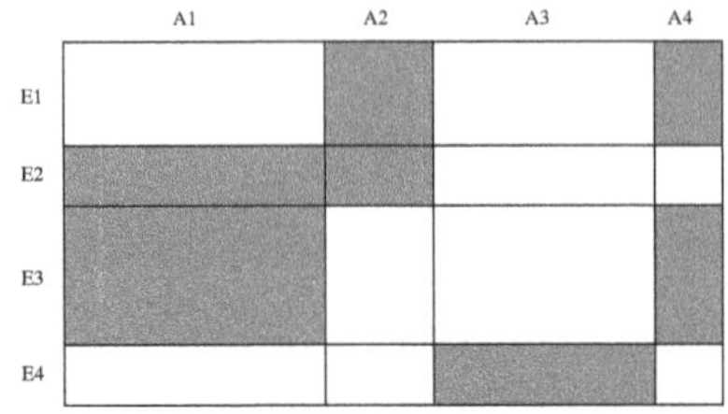


**Figure 3.1.** *Binary table reorganized according to row and column partitions*

## 3.1. Example and notation

This purpose can be shown precisely with a simple example. Let us consider data matrix x in Figure 3.2(a). The rows correspond to a set of 10 micro computers and the columns to 10 properties that these computers may (value 1) or may not have (value 0). With this example, if the partitions z and w are, respectively, $\{\{a,d,h\}, \{b,e,f,j\}, \{c,g,i\}\}$ and $\{\{1,3,5,8,10\}, \{2,4,6,7,9\}\}$, we obtain reorganized data matrix x (see Figure 3.2(b)) by reorganizing rows and columns according to these two partitions and the initial binary matrix can be summarized by the binary matrix a (see Figure 3.2(c)) when crossing the two partitions. In doing so, the 100 binary values of the initial data matrix x were summarized by the six binary values of the matrix a.

| | 1 | 2 | 3 | 4 | 5 | 6 | 7 | 8 | 9 | 10 |
|---|---|---|---|---|---|---|---|---|---|---|
| a | 1 | 0 | 1 | 0 | 1 | 0 | 0 | 1 | 0 | 1 |
| b | 0 | 1 | 0 | 1 | 0 | 1 | 1 | 0 | 1 | 0 |
| c | 1 | 0 | 0 | 0 | 0 | 0 | 0 | 1 | 1 | 0 |
| d | 1 | 0 | 1 | 0 | 0 | 0 | 0 | 1 | 0 | 0 |
| e | 0 | 1 | 0 | 1 | 0 | 1 | 1 | 0 | 1 | 0 |
| f | 0 | 1 | 0 | 0 | 0 | 1 | 1 | 0 | 1 | 0 |
| g | 0 | 1 | 0 | 0 | 0 | 0 | 0 | 1 | 0 | 1 |
| h | 1 | 0 | 1 | 0 | 1 | 1 | 0 | 1 | 1 | 1 |
| i | 1 | 0 | 0 | 1 | 0 | 0 | 0 | 0 | 0 | 1 |
| j | 0 | 1 | 0 | 1 | 0 | 0 | 1 | 0 | 0 | 0 |

(a) Binary data x

| | | 1 | | | | | 2 | | | | |
|---|---|---|---|---|---|---|---|---|---|---|---|
| | | 1 | 3 | 5 | 8 | 10 | 2 | 4 | 6 | 7 | 9 |
| | a | 1 | 1 | 1 | 1 | 1 | 0 | 0 | 0 | 0 | 0 |
| A | d | 1 | 1 | 0 | 1 | 0 | 0 | 0 | 0 | 0 | 0 |
| | h | 1 | 1 | 1 | 1 | 1 | 0 | 0 | 1 | 0 | 1 |
| | b | 0 | 0 | 0 | 0 | 0 | 1 | 1 | 1 | 1 | 1 |
| B | e | 0 | 0 | 0 | 0 | 0 | 1 | 1 | 1 | 1 | 1 |
| | f | 0 | 0 | 0 | 0 | 0 | 1 | 0 | 1 | 1 | 1 |
| | j | 0 | 0 | 0 | 0 | 0 | 1 | 1 | 0 | 1 | 0 |
| | c | 1 | 0 | 0 | 1 | 0 | 0 | 0 | 0 | 0 | 1 |
| C | g | 0 | 0 | 0 | 1 | 1 | 1 | 0 | 0 | 0 | 0 |
| | i | 1 | 0 | 0 | 0 | 1 | 0 | 1 | 0 | 0 | 0 |

(b) Reorganized data matrix x

| | 1 | 2 |
|---|---|---|
| A | 1 | 0 |
| B | 0 | 1 |
| C | 0 | 0 |

(c) Summary matrix a

**Figure 3.2.** *Microcomputer data*

To obtain a summary of initial data, we will see that the algorithms proposed are based, at each step, on intermediate matrices $\mathbf{x}^{\mathbf{z}}$, $\mathbf{x}^{\mathbf{w}}$ and $\mathbf{x}^{\mathbf{zw}}$ of reduced size defined in Table 3.1 whose values obtained with the previous example are given in Table 3.2.

| Matrix | Size | Definition |
|---|---|---|
| $\mathbf{x}^{\mathbf{z}} = (x^{\mathbf{z}}_{kj})$ | $(g \times d)$ | $x^{\mathbf{z}}_{kj} = \sum_i z_{ik} x_{ij}$ |
| $\mathbf{x}^{\mathbf{w}} = (x^{\mathbf{w}}_{i\ell})$ | $(n \times m)$ | $x^{\mathbf{w}}_{i\ell} = \sum_j w_{j\ell} x_{ij}$ |
| $\mathbf{x}^{\mathbf{zw}} = (x^{\mathbf{zw}}_{k\ell})$ | $(g \times m)$ | $x^{\mathbf{zw}}_{k\ell} = \sum_{i,j} z_{ik} w_{j\ell} x_{ij}$ |

**Table 3.1.** *Reduced matrices, sizes and definitions of* $\mathbf{x}^{\mathbf{z}}$, $\mathbf{x}^{\mathbf{w}}$ *and* $\mathbf{x}^{\mathbf{zw}}$

$$\mathbf{x}^{\mathbf{z}} = \begin{pmatrix} 3 & 3 & 2 & 3 & 2 & 0 & 0 & 1 & 0 & 1 \\ 0 & 0 & 0 & 0 & 0 & 4 & 3 & 3 & 4 & 3 \\ 2 & 0 & 0 & 2 & 2 & 1 & 1 & 0 & 0 & 1 \end{pmatrix}$$

$$\mathbf{x}^{\mathbf{w}} = \begin{pmatrix} 5 & 0 \\ 3 & 0 \\ 5 & 2 \\ 0 & 5 \\ 0 & 5 \\ 0 & 4 \\ 0 & 3 \\ 2 & 1 \\ 2 & 1 \\ 2 & 1 \end{pmatrix}$$

$$\mathbf{x}^{\mathbf{zw}} = \begin{pmatrix} 13 & 2 \\ 0 & 17 \\ 6 & 3 \end{pmatrix}$$

**Table 3.2.** *Intermediate data matrices* $\mathbf{x}^{\mathbf{z}}$, $\mathbf{x}^{\mathbf{w}}$ *and* $\mathbf{x}^{\mathbf{zw}}$ *obtained with the microcomputer data*

Moreover, we will also use the notation $x^{\mathbf{z}}_{k.} = \sum_i z_{ik}x_{i.}$ and $x^{\mathbf{w}}_{.\ell} = \sum_j w_{j\ell}x_{.j}$. In a similar way, we can define, with the fuzzy partitions $\widetilde{z}$ and $\widetilde{w}$, the matrices $\mathbf{x}^{\widetilde{\mathbf{z}}}$, $\mathbf{x}^{\widetilde{\mathbf{w}}}$ and $\mathbf{x}^{\widetilde{\mathbf{z}}\widetilde{\mathbf{w}}}$.

To obtain these homogeneous blocks (or biclusters), most methods used in this context can be grouped into two families: metric methods and model-based clustering. Sections 3.2 and 3.3 will be, respectively, devoted to these two kinds of approaches.

## 3.2. Metric approach

To obtain such homogeneous blocks, we must minimize the number of times where the value $x_{ij}$ is different from the value $a_{k\ell}$ associated with the clusters pair $(k,\ell)$ to which $(i,j)$ belongs. This quantity represents the difference between the initial data and the summary data. Then, the problem consists of optimizing the following objective function

$$\mathrm{W}(\mathbf{z},\mathbf{w},\mathbf{a}) = \sum_{i,j,k,\ell} z_{ik}w_{j\ell}|x_{ij} - a_{k\ell}|,$$

where $a_{k\ell} \in \{0,1\}$.

The minimization of this criterion leads to obtaining homogeneous blocks of 0 or 1 by reorganizing rows and columns according to the partitions $\mathbf{z}$ and $\mathbf{w}$. Hence, each block $k\ell$, defined by the elements $x_{ij}$ for $i$ belonging to the $k$th cluster and $j$ to the $\ell$th cluster, is characterized by $a_{k\ell}$, which is the highest frequency value. To interpret the results, some statistics can be computed to evaluate the quality of the partition into blocks. For instance, we can define the proportion of block $k\ell$ values equal to $a_{k\ell}$ and therefore measure the degree of homogeneity of this block.

When we only deal with the research of one partition, we obtain the well-known maximal predictive classification criterion proposed by Gower [GOW 74]. Let us mention that we also find this criterion when we use a classification maximum likelihood criterion [CEL 92] or an entropy criterion [GYL 94] on the Bernoulli mixture model.

Besides, if we constrain the summary to have a diagonal structure (same number of clusters for the two partitions, 1 on the diagonal and 0 everywhere else), we find the criterion proposed by Garcia and Proth [GAR 86]. Let us note that introducing such a constraint in the following algorithm is very easy.

In algorithmic terms, to minimize this objective function and using the strategy described in section 2.1 of Chapter 2, Govaert [GOV 83, GOV 95] proposed the following CROBIN algorithm 3.1.

---

**Algorithm 3.1** CROBIN (here $\lfloor x \rceil$ is the nearest integer function)

---

**input:** x, $g$, $m$
**initialization:** z, w, $a_{k\ell} = \lfloor \frac{x^{\mathbf{zw}}_{k\ell}}{z_{.k}w_{.\ell}} \rceil$
**repeat**
  $x^{\mathbf{w}}_{i\ell} = \sum_j w_{j\ell} x_{ij}$
  **repeat**
    **step 1.** $z_i = \arg\min_k \sum_\ell w_{j\ell} |x^{\mathbf{w}}_{i\ell} - w_{.\ell} a_{k\ell}|$
    **step 2.** $a_{k\ell} = \lfloor \frac{\sum_k z_{ik} x^{\mathbf{w}}_{i\ell}}{z_{.k}w_{.\ell}} \rceil$
  **until** convergence
  $x^{\mathbf{z}}_{kj} = \sum_i z_{ik} x_{ij}$
  **repeat**
    **step 3.** $w_j = \arg\min_\ell \sum_k z_{ik} |x^{\mathbf{z}}_{kj} - z_{.k} a_{k\ell}|$
    **step 4.** $a_{k\ell} = \lfloor \frac{\sum_j w_{j\ell} x^{\mathbf{z}}_{kj}}{z_{.k}w_{.\ell}} \rceil$
  **until** convergence
**until** convergence
**return** z, w, a

---

REMARK 3.1.– Steps 1 and 2 can be seen as a dynamic cluster algorithm [DID 79] applied to the $n \times m$ matrix $\mathbf{x}^{\mathbf{w}}$ with the $L_1$ distance and centroids

$$(w_{.1}a_{k1}, \ldots, w_{.m}a_{km})^t,$$

which minimizes the criterion $W(\mathbf{z}, \mathbf{a}|\mathbf{w}) = \sum_{i,k} z_{ik} \sum_\ell |x^{\mathbf{w}}_{i\ell} - w_{.\ell}a_{k\ell}|$. Similarly, steps 3 and 4 can be seen as a dynamic cluster algorithm applied to the $d \times n$ matrix $\mathbf{x}^{\mathbf{z}}$ with the $L_1$ distance and centroids

$$(z_{.1}a_{1\ell}, \ldots, z_{.g}a_{g\ell})^t,$$

which minimizes the criterion $W(\mathbf{w}, \mathbf{a}|\mathbf{z}) = \sum_{j,\ell} w_{j\ell} \sum_k |x^{\mathbf{z}}_{kj} - z_{.k}a_{k\ell}|$.

The structure of this program allows us to control the size of the matrices used in each step making it possible to use this program for large, and also to take into account sparse, data. This particular structure is an important feature in all of the programs described in this book.

## 3.3. Bernoulli latent block model and algorithms

### 3.3.1. *The model*

Using the latent block model (LBM) described in section 2.3 of Chapter 2, the binary data situation leads us to assume that for each block $k\ell$, the values $x_{ij}$ are distributed according to the Bernoulli distribution $\mathcal{B}(\alpha_{k\ell})$ ($\alpha_{k\ell} \in \mathbb{R}$ and $0 < \alpha_{k\ell} < 1$) for which the pdf is

$$f(x_{ij}; \alpha_{k\ell}) = (\alpha_{k\ell})^{x_{ij}}(1 - \alpha_{k\ell})^{(1-x_{ij})}.$$

Denoting $\boldsymbol{\theta} = (\boldsymbol{\pi}, \boldsymbol{\rho}, \boldsymbol{\alpha})$ where $\boldsymbol{\pi} = (\pi_1, \ldots, \pi_g)$, $\boldsymbol{\rho} = (\rho_1, \ldots, \rho_m)$ and $\boldsymbol{\alpha} = (\alpha_{11}, \ldots, \alpha_{gm})$, we obtain the following pdf according to Govaert and Nadif [GOV 03]

$$f(\mathbf{x};\boldsymbol{\theta}) = \sum_{(\mathbf{z},\mathbf{w})\in\mathcal{Z}\times\mathcal{W}} \prod_{i,k} \pi_k^{z_{ik}} \prod_{j,\ell} \rho_\ell^{w_{j\ell}}$$

$$\times \prod_{i,j,k,\ell} \left( (\alpha_{k\ell})^{x_{ij}} (1-\alpha_{k\ell})^{(1-x_{ij})} \right)^{z_{ik}w_{j\ell}},$$

the complete-data log-likelihood becomes

$$\mathrm{L_C}(\mathbf{z},\mathbf{w},\boldsymbol{\theta}) = \sum_{i,k} z_{ik} \log \pi_k + \sum_{j,\ell} w_{j\ell} \log \rho_\ell$$

$$+ \sum_{i,k,j,\ell} z_{ik} w_{j\ell} x_{ij} \log \frac{\alpha_{k\ell}}{1-\alpha_{k\ell}} + \sum_{k,\ell} z_{k.} w_{.\ell} \log(1-\alpha_{k\ell}),$$

and then, the following fuzzy clustering criterion can be deduced

$$\mathrm{F_C}(\widetilde{\mathbf{z}},\widetilde{\mathbf{w}};\boldsymbol{\theta}) = \mathrm{L_C}(\widetilde{\mathbf{z}},\widetilde{\mathbf{w}},\boldsymbol{\theta}) + \mathrm{H}(\widetilde{\mathbf{z}}) + \mathrm{H}(\widetilde{\mathbf{w}}). \qquad [3.1]$$

### 3.3.2. *Model identifiability*

It is well-known that simple multivariate Bernoulli mixtures are not identifiable [GYL 94], regardless of the invariance to relabeling. Allman *et al.* [ALL 09] set a sufficient condition to their identifiability that cannot be applied to LBM. The following theorem on sufficient conditions ensuring the identifiability of the Bernoulli LBM can be established [KER 13].

THEOREM 3.1.– With $\alpha$, the matrix of the Bernoulli coefficients, $\pi$ and $\rho$, the row and column mixing proportions of the mixture, assume that $n \geq 2m-1$ and $d \geq 2g-1$ and

– ($H_1$) for all $1 \leq k \leq g$, $\pi_k > 0$ and the coordinates of vector $\tau = \alpha\rho$ are distinct;

– ($H_2$) for all $1 \leq \ell \leq m$, $\rho_\ell > 0$ and the coordinates of vector $\sigma = \pi^t\alpha$ are distinct (where $\pi^t$ is the transpose of $\pi$);

then, the binary LBM is identifiable.

$H_1$ and $H_2$ are not strongly restrictive since the set of vectors $\tau$ and $\sigma$ violating them is of Lebesgue measure 0. Therefore, theorem 3.1 asserts the generic identifiability of the Bernoulli LBM, which is a "practical" identifiability, explaining why it works in the applications [CAR 00]. These assumptions could appear to be somewhat unnatural; however, (1) it is not surprising that the number of row labels $g$ (respectively, column labels $m$) is constrained by the number of columns $d$ (respectively, rows $n$). In case of a simple finite mixture of $g$ different Bernoulli products with $d$ components, the more clusters you define in the mixture, the more components for the multivariate Bernoulli you need in order to ensure the identifiability: see also the following condition $d > 2\lceil \log_2 g \rceil + 1$ of Allman *et al.* [ALL 09] for simple Bernoulli mixtures, where $\lceil . \rceil$ is the ceil function. (2) Assumptions $H_1$ and $H_2$ are the extensions of an assumption made to ensure the identifiability of the stochastic block model [CEL 12], where $n = d$ and $\mathbf{z} = \mathbf{w}$.

### 3.3.3. *Binary* LBVEM *and* LBCEM *algorithms*

Taking into account the definition of the pdf $f(x_{ij}; \alpha_{k\ell})$, the variational approximation of the expectation–maximization (EM) algorithm described in section 2.4.1 of Chapter 2 and maximizing $\mathrm{F_C}(\widetilde{\mathbf{z}}, \widetilde{\mathbf{w}}; \boldsymbol{\theta})$ can be completed by the computation of the block parameters $\alpha_{k\ell}$s. We obtain for all $k, \ell$

$$\begin{aligned}\alpha_{k\ell} &= \arg\max_{\alpha_{k\ell}} \sum_{i,j} \widetilde{z}_{ik}\widetilde{w}_{j\ell} \log f(x_{ij}, \boldsymbol{\alpha}_{k\ell}) \\ &= \arg\max_{\alpha_{k\ell}} \sum_{i,j} \widetilde{z}_{ik}\widetilde{w}_{j\ell} \left(x_{ij} \log \alpha_{k\ell} + (1 - x_{ij}) \log (1 - \alpha_{k\ell})\right) \\ &= \arg\max_{\alpha_{k\ell}} \left(x_{k\ell}^{\widetilde{\mathbf{z}}\widetilde{\mathbf{w}}} \log \alpha_{k\ell} + (\widetilde{z}_{.k}\widetilde{w}_{.\ell} - x_{k\ell}^{\widetilde{\mathbf{z}}\widetilde{\mathbf{w}}}) \log (1 - \alpha_{k\ell})\right),\end{aligned}$$

which yields $\alpha_{k\ell} = \frac{x_{k\ell}^{\widetilde{\mathbf{z}}\widetilde{\mathbf{w}}}}{\widetilde{z}_{.k}\widetilde{w}_{.\ell}}$ and finally algorithm 3.2.

**Algorithm 3.2** Bernoulli LBVEM

**input:** x, $g$, $m$
**initialization:** $\widetilde{\mathrm{z}}$, $\widetilde{\mathrm{w}}$, $\pi_k = \frac{\widetilde{z}_{.k}}{n}$, $\rho_\ell = \frac{\widetilde{w}_{.\ell}}{d}$, $\alpha_{k\ell} = \frac{x_{k\ell}^{\widetilde{\mathbf{z}}\widetilde{\mathbf{w}}}}{\widetilde{z}_{.k}\widetilde{w}_{.\ell}}$
**repeat**
  $x_{i\ell}^{\widetilde{\mathbf{w}}} = \sum_j \widetilde{w}_{j\ell} x_{ij}$
  **repeat**
    **step 1.** $\widetilde{z}_{ik} \propto \pi_k \exp \sum_\ell \left(x_{i\ell}^{\widetilde{\mathbf{w}}} \ln \frac{\alpha_{k\ell}}{1-\alpha_{k\ell}} + \widetilde{w}_{.\ell} \ln(1 - \alpha_{k\ell})\right)$
    **step 2.** $\pi_k = \frac{\widetilde{z}_{.k}}{n}$, $\alpha_{k\ell} = \frac{\sum_i \widetilde{z}_{ik} x_{i\ell}^{\widetilde{\mathbf{w}}}}{\widetilde{z}_{.k}\widetilde{w}_{.\ell}}$
  **until** convergence
  $x_{kj}^{\widetilde{\mathbf{z}}} = \sum_i \widetilde{z}_{ik} x_{ij}$
  **repeat**
    **step 3.** $\widetilde{w}_{j\ell} \propto \rho_\ell \exp \sum_k \left(x_{jk}^{\widetilde{\mathbf{z}}} \ln \frac{\alpha_{k\ell}}{1-\alpha_{k\ell}} + \widetilde{z}_{.k} \ln(1 - \alpha_{k\ell})\right)$
    **step 4.** $\rho_\ell = \frac{\widetilde{w}_{.\ell}}{d}$, $\alpha_{k\ell} = \frac{\sum_j \widetilde{w}_{j\ell} x_{kj}^{\widetilde{\mathbf{z}}}}{\widetilde{z}_{.k}\widetilde{w}_{.\ell}}$
  **until** convergence
**until** convergence
**return** $\boldsymbol{\pi}$, $\boldsymbol{\rho}$, $\boldsymbol{\alpha}$

Knowing that we have the following two relations shown in [GOV 08]

$$\mathrm{F_C}(\widetilde{\mathbf{z}}, \widetilde{\mathbf{w}}, \boldsymbol{\theta}) = \mathrm{F_C}(\widetilde{\mathbf{z}}, \boldsymbol{\alpha}, \boldsymbol{\pi}|\widetilde{\mathbf{w}}, \boldsymbol{\rho}) + \sum_{j,\ell} \widetilde{w}_{j\ell} \log \rho_\ell + \mathrm{H}(\widetilde{\mathbf{w}})$$

and

$$\mathrm{F_C}(\widetilde{\mathbf{z}}, \widetilde{\mathbf{w}}, \boldsymbol{\theta}) = \mathrm{F_C}(\widetilde{\mathbf{w}}, \boldsymbol{\alpha}, \boldsymbol{\rho}|\widetilde{\mathbf{z}}, \boldsymbol{\pi}) + \sum_{i,k} \widetilde{z}_{j\ell} \log \pi_k + \mathrm{H}(\widetilde{\mathbf{z}}),$$

where $\mathrm{F_C}(\widetilde{\mathbf{z}}, \boldsymbol{\alpha}, \boldsymbol{\pi}|\widetilde{\mathbf{w}}, \boldsymbol{\rho})$ and $\mathrm{F_C}(\widetilde{\mathbf{w}}, \boldsymbol{\alpha}, \boldsymbol{\rho}|\widetilde{\mathbf{z}}, \boldsymbol{\pi})$ correspond, respectively, to Hathaway's criteria defined in section 1.4.5 of Chapter 1 for the row clustering of the data matrix $\mathrm{x}^{\mathrm{w}}$ and for the column clustering of the data matrix $\mathrm{x}^{\mathrm{z}}$, steps 1 and 2 can be seen as steps of EM applied on the $n \times m$ matrix $\mathrm{x}^{\widetilde{\mathrm{w}}}$ and maximizing $\mathrm{F_C}(\widetilde{\mathbf{z}}, \boldsymbol{\alpha}, \boldsymbol{\pi}|\widetilde{\mathbf{w}}, \boldsymbol{\rho})$. Similarly, steps 3 and 4 can be seen as steps of EM applied on the $g \times d$ matrix $\mathrm{x}^{\widetilde{\mathbf{z}}}$ and maximizing $\mathrm{F_C}(\widetilde{\mathbf{w}}, \boldsymbol{\alpha}, \boldsymbol{\rho}|\widetilde{\mathbf{z}}, \boldsymbol{\pi})$.

In a similar way, the maximization of the complete-data log-likelihood $\mathrm{L_C}$ can be performed by the alternated maximization of conditional complete-data log-likelihoods $\mathrm{L_C}(\mathbf{z}, \boldsymbol{\alpha}, \boldsymbol{\pi}|\mathbf{w}, \boldsymbol{\rho})$ and $\mathrm{L_C}(\mathbf{w}, \boldsymbol{\alpha}, \boldsymbol{\rho}|\mathbf{z}, \boldsymbol{\pi})$ which correspond, respectively, to the complete-data log-likelihoods for the row clustering of the data matrix $\mathrm{x}^{\mathrm{w}}$ and for the column clustering of the data matrix $\mathrm{x}^{\mathrm{z}}$. Then, the principal steps of the algorithm used for this task are reported in algorithm 3.3.

Different comparisons were performed by Govaert and Nadif [GOV 05, GOV 08] confirming that LBVEM outperforms LBCEM in terms of estimation. In Table 3.3, we illustrate this observation by reporting a comparison between the two algorithms LBCEM and LBVEM on $200 \times 120$ data matrix. The quality of estimate measured by the deviance between the estimated parameter $\boldsymbol{\theta}$ and the true parameter $\boldsymbol{\theta}^0$, whose direct calculation is not possible, is approximated by $D^2(\boldsymbol{\theta}, \boldsymbol{\theta}^0)$ ($D^2$ denotes the Euclidean distance) expressed by the following formula

$$D^2(\boldsymbol{\theta}, \boldsymbol{\theta}^0) = \sum_{k,\ell} (\alpha_{k\ell} - \alpha^0_{k\ell})^2 + \sum_{k} (\pi_k - \pi^0_k)^2 + \sum_{\ell} (\rho_\ell - \rho^0_\ell)^2.$$

We observe that the estimated parameters are closer to the true values with LBVEM than with LBCEM. The value of $D^2(\boldsymbol{\theta}, \boldsymbol{\theta}^0)$ is equal to 0.0252 for LBVEM and is equal to 0.0824 for LBCEM. In terms of co-clustering, in [GOV 05, GOV 08], the authors showed that LBVEM outperforms LBCEM even when the clusters are well separated. However, the latter is more scalable and it converges within 30 iterations.

**Algorithm 3.3** Bernoulli LBCEM

**input:** x, $g$, $m$
**initialization:** z, w, $\pi_k = \frac{z_{.k}}{n}$, $\rho_\ell = \frac{w_{.\ell}}{d}$, $\alpha_{k\ell} = \frac{x_{k\ell}^{\tilde{\mathbf{z}}\tilde{\mathbf{w}}}}{\tilde{z}_{.k}\tilde{w}_{.\ell}}$
**repeat**
  $x_{i\ell}^{\mathbf{w}} = \sum_j w_{j\ell} x_{ij}$
  **repeat**
    **step 1.** $z_i = \arg\max_k \{\sum_\ell \{x_{i\ell}^{\mathbf{w}} \ln \frac{\alpha_{k\ell}}{1-\alpha_{k\ell}} + w_{.\ell} \ln(1-\alpha_{k\ell})\} + \ln \pi_k\}$
    **step 2.** $\pi_k = \frac{z_{.k}}{n}$, $\alpha_{k\ell} = \frac{\sum_i z_{ik} x_{i\ell}^{\mathbf{w}}}{z_{.k} w_{.\ell}}$
  **until** convergence
  $x_{kj}^{\mathbf{z}} = \sum_i z_{ik} x_{ij}$
  **repeat**
    **step 3.** $w_j = \arg\max_\ell \{\sum_k \{x_{kj}^{\mathbf{z}} \ln \frac{\alpha_{k\ell}}{1-\alpha_{k\ell}} + z_{.k} \ln(1-\alpha_{k\ell})\} + \ln \rho_\ell\}$
    **step 4.** $\rho_\ell = \frac{w_{.\ell}}{d}$, $\alpha_{k\ell} = \frac{\sum_j w_{j\ell} x_{kj}^{\mathbf{z}}}{z_{.k} w_{.\ell}}$
  **until** convergence
**until** convergence
**return** z, w, $\pi$, $\rho$, $\alpha$

Furthermore, the complexity of these algorithms depends on the average time of execution of an iteration. The number of iterations, depends on the degree of mixture, the size of the data and the complexity of the model. For the average time of an iteration, we checked on many examples that this one was independent of the degree of the mixture and that it depends, in a roughly linear way, on the sizes $n$ and $d$ of the data and the number $g$ and $m$ of clusters.

| $\boldsymbol{\theta}$ | True values ($\boldsymbol{\theta}^0$) | Parameter estimations values by LBVEM | Parameter estimations by LBCEM |
|---|---|---|---|
| $\pi_1$ | 0.2 | 0.1979 | 0.1900 |
| $\pi_2$ | 0.3 | 0.3140 | 0.3400 |
| $\pi_3$ | 0.5 | 0.4881 | 0.4700 |
| $\rho_1$ | 0.3 | 0.2929 | 0.2583 |
| $\rho_2$ | 0.7 | 0.7071 | 0.7417 |
| $\alpha$ | $\begin{pmatrix} 0.60 & 0.40 \\ 0.40 & 0.60 \\ 0.60 & 0.65 \end{pmatrix}$ | $\begin{pmatrix} 0.6067 & 0.4026 \\ 0.4089 & 0.6041 \\ 0.5989 & 0.6565 \end{pmatrix}$ | $\begin{pmatrix} 0.6188 & 0.4063 \\ 0.3861 & 0.6000 \\ 0.6095 & 0.6559 \end{pmatrix}$ |
| $D^2(\boldsymbol{\theta}, \boldsymbol{\theta}^0)$ | 0 | **0.0252** | 0.0824 |

**Table 3.3.** *True values and their estimates by* LBVEM *and* LBCEM *for* $n \times d = 200 \times 120$ *data matrix*

## 3.4. Parsimonious Bernoulli LBMs

As in section 1.7.2 of Chapter 1, a parsimonious model can be defined by breaking down the block parameter into a "center" parameter and a "dispersion" parameter and by imposing constraints on the dispersion parameter. More precisely, each parameter $\alpha_{k\ell}$ is replaced by $a_{k\ell} \in \{0,1\}$ and $\varepsilon_{k\ell} \in [0, 1/2]$ with

$$\begin{cases} a_{k\ell} = 1 \text{ and } \varepsilon_{k\ell} = 1 - \alpha_{k\ell} \text{ if } \alpha_{k\ell} \in [1/2, 1] \\ a_{k\ell} = 0 \text{ and } \varepsilon_{k\ell} = \alpha_{k\ell} \text{ if } \alpha_{k\ell} \in [0, 1/2[. \end{cases}$$

Thus, the Bernoulli pdf can be written as

$$f(x_{ij}; (a_{k\ell}, \varepsilon_{k\ell})) = (\varepsilon_{k\ell})^{|x_{ij} - a_{k\ell}|}(1 - \varepsilon_{k\ell})^{1 - |x_{ij} - a_{k\ell}|},$$

where

– $a_{k\ell} \in \{0,1\}$, which is the most frequent binary value of the block, represents the center of the block, and;

– $\varepsilon_{k\ell} \in ]0, 1/2[^d$, which is the probability of any particular variable having a value different from that of the center of the block, represents the dispersion of the component.

Starting from this formulation, we arrive at parsimonious situations by stipulating certain constraints: so, the $[\varepsilon]$ model is defined by stipulating that the dispersion should not depend either on the row cluster or on the column cluster, the $[\varepsilon_k]$ model by stipulating that it should depend only on the row cluster and the $[\varepsilon^j]$ model by stipulating that it should depend only on the column cluster.

This parsimonious parameterization allows us to study the relationship between the CROBIN algorithm described in section 3.2 and the LBM. Indeed, in the simplest case, in the $[\varepsilon]$ model, given identical proportions ($\pi_k = 1/g$ and $\rho_\ell = 1/m$), the complete-data log-likelihood is equal to

$$\log \frac{\varepsilon}{1-\varepsilon} \mathrm{W}(\mathbf{z}, \mathbf{w}, \mathbf{a}) + nd \log(1-\varepsilon) - n \log g - d \log m,$$

where $\mathrm{W}(\mathbf{z}, \mathrm{w}, \mathrm{a}) = \sum_{i,j,k,\ell} z_{ik} w_{j\ell} |x_{ij} - a_{k\ell}|$ is the criterion optimized by the algorithm CROBIN described in section 3.2 so that maximizing $\mathrm{L_C}(\mathbf{z}, \mathrm{w}, \boldsymbol{\theta})$ is equivalent to minimizing $\mathrm{W}(\mathbf{z}, \mathrm{w}, \mathrm{a})$. In addition, it clearly appears that the steps maximizing alternatively $\mathrm{L_C}(\mathbf{z}, \boldsymbol{\alpha}, \boldsymbol{\pi}|\mathrm{w}, \boldsymbol{\rho})$ and $\mathrm{L_C}(\mathrm{w}, \boldsymbol{\alpha}, \boldsymbol{\rho}|\mathbf{z}, \boldsymbol{\pi})$ correspond to steps of CROBIN minimizing $\mathrm{W}(\mathbf{z}, \mathrm{a}|\mathrm{w})$ and $\mathrm{W}(\mathrm{w}, \mathrm{a}|\mathbf{z})$.

### 3.5. Categorical data

Developments presented in the first part of this chapter can be extended to a population of $n$ observations described with $d$ categorical variables of the same nature with $r$ levels being available. Saying that the categorical variables are of the same nature means that it is possible to code them in the same (and natural) way. This assumption is needed to ensure that decomposing the data set in a block structure makes sense. Let $\mathrm{x} = (x_{ij}, i = 1, \ldots, n; j = 1, \ldots, d)$ be the data matrix defined on $I \times J$ where $x_{ij} = h$, $1 \leq h \leq r$, $r$ is the number of levels of the $J$ variables. In the following, an alternative representation of the data set with binary indicator vectors will often be used for mathematical convenience: $\mathrm{x}_{ij} = (x_{ij}^h, h = 1, \ldots, r)$ with $x_{ij}^h = 1$ when $x_{ij} = h$ and $x_{ij}^h = 0$.

To extend results on binary data to categorical data, a categorical LBM is considered: the conditional distribution of the outcome $x_{ij}$ knowing the labels $z_{ik}$ and $w_{j\ell}$ is a categorical distribution $\mathcal{C}(\alpha_{k\ell})$, of parameter $\alpha_{k\ell} = (a_{k\ell}^h)_{h=1,\ldots,r}$, where $a_{k\ell}^h \in (0;1)$ and $\sum_h a_{k\ell}^h = 1$. Using the binary indicator vector $\mathrm{x}_{ij}$, the pdf per block is

$$f(\mathrm{x}_{ij}; \alpha_{k\ell}) = \prod_h (a_{k\ell}^h)^{x_{ij}^h}$$

and the mixture pdf is

$$f(\mathrm{x}; \boldsymbol{\pi}, \boldsymbol{\rho}, \boldsymbol{\alpha}) = \sum_{\mathbf{z},\mathbf{w}} \prod_{i,k} \pi_k^{z_{ik}} \prod_{j,\ell} \rho_\ell^{w_{j\ell}} \prod_{i,j,k,\ell,h} (a_{k\ell}^h)^{z_{ik} w_{j\ell} x_{ij}^h}.$$

The $g + m + (r-1)gm - 2$ parameters to be estimated are $\boldsymbol{\pi}$, $\boldsymbol{\rho}$ and $\boldsymbol{\alpha}$.

For the identifiability, theorem 3.1 is easily extended to the categorical case, where the assumptions are defined on vectors $\boldsymbol{\tau}^h = \boldsymbol{\alpha}^h \boldsymbol{\rho}$ and $\boldsymbol{\sigma}^h = \boldsymbol{\pi}^t \boldsymbol{\alpha}^h$, with $h = 1, \ldots, r$ and $\boldsymbol{\alpha}^h = (\alpha_{k\ell}^h)_{k=1,\ldots,g;\ell=1,\ldots,m}$.

The algorithms 3.2 and 3.3 can be easily extended by replacing the computation of $\alpha_{k\ell}$ by the following computations of $a_{k\ell}^h$

$$a_{k\ell}^h = \frac{\sum_{i,j} \widetilde{z}_{ik} \widetilde{\mathrm{w}}_{j\ell} x_{ij}^h}{\widetilde{z}_{.k} \widetilde{\mathrm{w}}_{.\ell}}$$

in the variational EM (VEM) situation and

$$a_{k\ell}^h = \frac{\sum_{i,j} z_{ik} w_{j\ell} x_{ij}^h}{z_{.k} w_{.\ell}}$$

in the classification EM (CEM) situation.

## 3.6. Bayesian inference

Bayesian inference in statistics can be regarded as a well-grounded tool for regularizing maximum likelihood estimates in a poorly posed setting. In the LBM setting, Bayesian inference could be thought of as useful for avoiding spurious solutions and thus attenuating the "empty cluster" problem. In particular, for the categorical LBM, it is possible to consider non-informative prior distribution for the model parameters (see Figure 3.3)

$$\boldsymbol{\pi} \sim \mathcal{D}(a, \ldots, a), \ \boldsymbol{\rho} \sim \mathcal{D}(a, \ldots, a), \ \boldsymbol{\alpha} \sim \prod_{k,\ell} \mathcal{D}(b, \ldots, b),$$

$\mathcal{D}(v, \ldots, v)$ denoting a Dirichlet distribution with parameter $v$. Obviously, since Dirichlet prior distributions are conjugate priors for the multinomial distribution, full conditional posterior distributions of the LBM parameters are closed form and Gibbs sampling is easy to implement.

Using Bayesian inference from a regularization perspective, the model parameter may be estimated by maximizing the posterior probability $p(\boldsymbol{\theta}|\mathrm{x})$ that leads to the so-called maximum *a posteriori* (MAP) estimate

$$\widehat{\boldsymbol{\theta}}_{MAP} = \arg\max_{\boldsymbol{\theta}} p(\boldsymbol{\theta}|\mathbf{x}).$$

The Bayes formula

$$\log p(\boldsymbol{\theta}|\mathbf{x}) = \log p(\mathbf{x}|\boldsymbol{\theta}) + \log p(\boldsymbol{\theta}) - \log p(\mathbf{x})$$

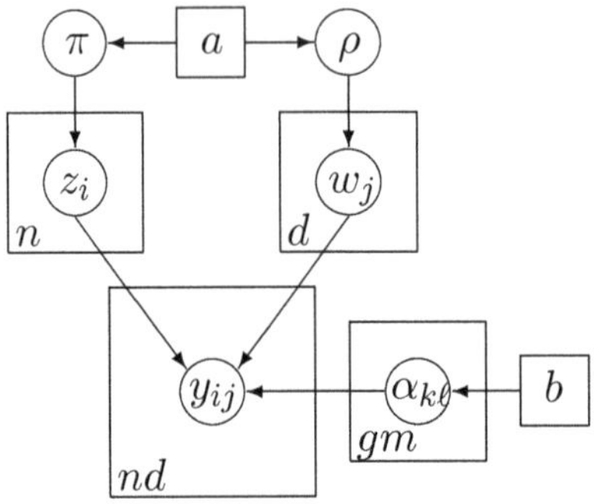


**Figure 3.3.** *Bayesian latent block model*

allows us to straightforwardly define an EM algorithm for the computation of the MAP estimate:

– The E step relies on the computation of the conditional expectation of the complete log-likelihood $Q(\boldsymbol{\theta}, \boldsymbol{\theta}')$ as for the maximum likelihood estimator

$$Q(\boldsymbol{\theta}, \boldsymbol{\theta}') = \mathbb{E}\big(\log p(\mathbf{x}, \mathbf{z}, \mathrm{w}|\boldsymbol{\theta}')|\mathrm{x}, \boldsymbol{\theta}'\big).$$

– The M step differs in that the objective function for the maximization process is equal to the $Q(\boldsymbol{\theta}, \boldsymbol{\theta}')$ function augmented by the logarithm of the prior probability

$$\boldsymbol{\theta} = \arg\max_{\boldsymbol{\theta}} \big(Q(\boldsymbol{\theta}, \boldsymbol{\theta}') + \log p(\boldsymbol{\theta})\big).$$

This M step forces an increase in the log posterior function $p(\boldsymbol{\theta}|\mathrm{x})$ [MCL 07, Chapter 5, p. 231]. For LBM, the Bayesian approach combined with the variational approximation leads to the V-Bayes algorithm, with the following M step:

*M step* (V-Bayes algorithm: estimation of the posterior mode)

$$\pi_k = \frac{a - 1 + \widetilde{z}_{.k}}{n + g(a-1)}, \quad \rho_\ell = \frac{a - 1 + \widetilde{\mathrm{w}}_{.\ell}}{d + m(a-1)},$$

$$a_{k\ell}^h = \frac{b - 1 + \sum_{i,j} \widetilde{z}_{ik} \widetilde{\mathrm{w}}_{j\ell}\, x_{ij}^h}{r(b-1) + \widetilde{z}_{.k} \widetilde{\mathrm{w}}_{.\ell}}.$$

The hyperparameters $a$ and $b$ are acting as regularization parameters. From this perspective, the choices of $a$ and $b$ are important. It clearly appears from the updating equations of this M step that V-Bayes is useless for uniform prior ($a = b = 1$) and worse than useless for Jeffreys prior ($a = b = 1/2$). Frühwirth-Schnatter [FRÜ 11] showed the great influence of $a$ for Bayesian inference in the Gaussian mixture context. Based on an asymptotic analysis by Rousseau and Mergensen [ROU 11] and on a thorough finite sample analysis, they advocated taking $a = 4$ for moderate dimensions ($g < 8$) and $a = 16$ for larger dimensions to avoid empty clusters.

However, as VEM, a V-Bayes algorithm could be expected to be highly dependent on its initial values. Thus, it could be of significance to initiate V-Bayes with the solution derived from the following Gibbs sampler:

1) Simulation of z according to $p(\mathbf{z}|\mathbf{x}, \mathbf{w}; \boldsymbol{\theta})$ as in SEM-Gibbs.

2) Simulation of w according to $p(\mathbf{w}|\mathbf{x}, \mathbf{z}; \boldsymbol{\theta})$ as in SEM-Gibbs.

3) Simulation of $\pi$ according to $\mathcal{D}(a + z_{.1}, \ldots, a + z_{.g})$.

4) Simulation of $\rho$ according to $\mathcal{D}(a + w_{.1}, \ldots, a + w_{.m})$.

5) Simulation of $\alpha_{k\ell}$ according to a $\mathcal{D}(b + x^1_{k\ell}, \ldots, b + x^r_{k\ell})$ for $k = 1, \ldots, g; \ell = 1, \ldots, m$ and with $x^h_{k\ell} = \sum_{i,j} z_{ik} w_{j\ell} x^h_{ij}$.

As Gibbs sampling explores the whole distribution, it is subject to label switching. This problem can be sensibly solved for the categorical LBM by using the identifiability conditions of theorem 3.1: these conditions define a natural order of row and column labels except on a set of parameters of measure zero. Hence, column (respectively, row) labels are reordered according to the ascending values of $\tau^h = \alpha^h \rho$

(respectively, $\sigma^h = \pi^t \alpha^h$) coordinates for a given $h$, after each steps 3 and 5 (respectively, 4 and 5).

## 3.7. Model selection

Choosing relevant numbers of clusters in an LBM is obviously of crucial importance. In the previous chapter, we mentioned the reasons that pose difficulties to dealing with the selection problem. However, it is possible to compute the exact integrated completed log-likelihood (ICL) of the categorical LBM and an asymptotic approximation of the integrated likelihood could be derived from this ICL criterion.

### 3.7.1. *The integrated completed log-likelihood (ICL)*

The ICL [BIE 00] is the logarithm of the integrated completed likelihood, which takes the following form

$$p(\mathbf{x}, \mathbf{z}, \mathbf{w} \mid g, m) = \int_{\Theta_{g,m}} p(\mathbf{x}, \mathbf{z}, \mathbf{w} \mid \boldsymbol{\theta}_{g,m}) p(\boldsymbol{\theta}_{g,m}) d\boldsymbol{\theta}_{g,m}$$

with

$$p(\mathbf{x}, \mathbf{z}, \mathbf{w} \mid \boldsymbol{\theta}_{g,m}) = \sum_{(\mathbf{z},\mathbf{w}) \in \mathcal{Z} \times \mathcal{W}} p(\mathbf{z}) p(\mathbf{w}) f(\mathbf{x}|\mathbf{z}, \mathbf{w}; \boldsymbol{\theta}_{g,m}),$$

where $f(\mathbf{x}|\mathbf{z}, \mathbf{w}; \boldsymbol{\theta}_{g,m}) = \prod_{i,j,k,\ell} f(x_{ij}; \alpha_{k\ell})^{z_{ik} w_{j\ell}}$, $\Theta_{g,m}$ being the parameter space of the model with $g \times m$ blocks and $p(\boldsymbol{\theta}_{g,m})$ a non-informative or a weakly informative prior distribution on $\boldsymbol{\theta}$. By taking into account the missing data, ICL is focusing on the clustering view of the model. For this very reason, ICL could be expected to select a stable model allowing us to partition the data with the greatest evidence [BIE 10].

For the categorical LBM, proper non-informative priors are available and ICL can be computed without requiring asymptotic approximations. Using the conjugacy properties of the prior Dirichlet distributions and the conditional independence of the $x_{ij}$ knowing the latent vectors z and w and the LBM parameters, it can be shown (see [KER 13]) that

$$\begin{aligned}\mathrm{ICL}(g,m) = &\log\Gamma(ga) + \log\Gamma(ma) - (m+g)\log\Gamma(a)\\ &+mg(\log\Gamma(rb) - r\log\Gamma(b))\\ &-\log\Gamma(n+ga) - \log\Gamma(d+ma)\\ &+\sum_k \log\Gamma(z_{.k}+a) + \sum_\ell \log\Gamma(w_{.\ell}+a)\\ &+\sum_{k,\ell}\left[\Big(\sum_h \log\Gamma(N_{k\ell}^h+b)\Big) - \log\Gamma(z_{.k}w_{.\ell}+rb)\right].\end{aligned}$$

In practice, the missing labels $\mathbf{z}, \mathbf{w}$ are to be chosen. Following [BIE 00], they are replaced by

$$(\hat{\mathbf{z}}, \hat{\mathbf{w}}) = \arg\max_{(\mathbf{z},\mathbf{w})} p(\mathbf{z}, \mathbf{w}|\mathbf{x}; \hat{\boldsymbol{\theta}}),$$

$\hat{\theta}$ being the maximum likelihood estimate of the LBM parameter derived from the SEM-Gibbs algorithm followed by the VEM algorithm as described in section 3.3.3.

### 3.7.2. *Penalized information criteria*

It could be of significance to analyze the behavior of standard information criteria as Bayesian information criterion (BIC) and $\mathrm{ICL}_{\mathrm{BIC}}$ derived from asymptotic approximations. Using the Stirling formula

$$\Gamma(z) \underset{+\infty}{\sim} z^{z-1/2}e^{-z}\sqrt{2\pi},$$

the asymptotic approximation of ICL as $n$ and $d$ tend to infinity is

$$\mathrm{ICL_{BIC}}(g,m) = \max_{\boldsymbol{\theta}} \log p(\mathbf{x}, \widehat{\mathbf{z}}, \widehat{\mathbf{w}}; \boldsymbol{\theta}) - \frac{g-1}{2}\log n - \frac{m-1}{2}\log d - \frac{gm(r-1)}{2}\log(nd).$$

This result is a generalization of the categorical case of the result proved in [KER 13] for the binary LBM. Moreover, BIC can be conjectured to have the following expression

$$\mathrm{BIC}(g,m) = \max_{\boldsymbol{\theta}} \log p(\mathbf{x}; \boldsymbol{\theta}) - \frac{g-1}{2}\log n - \frac{m-1}{2}\log d - \frac{gm(r-1)}{2}\log(nd).$$

As the maximized likelihood is unavailable for the LBM, it could be approximated by the maximized fuzzy function $\mathrm{F_C}$. But there is no reason to replace any criterion when available with its asymptotic approximation. Thus, ICL can be preferred to $\mathrm{ICL_{BIC}}$ or BIC.

## 3.8. Illustrative experiments

### 3.8.1. *Townships*

Hereafter, we illustrate the co-clustering aim by an example which consists of the characteristics (rows) of 16 townships {A,...,P} (columns), each cell indicating the presence 1 or absence 0 of a characteristic of a township (Table 3.4). The set of rows consists of the following characteristics: High School (hsco), Agricult Coop (agri), Railway Station (rail), One Room School (osco), Veterinary (vete), No Doctor (nodo), No Water Supply (nwat), Police Station (poli) and Land Reallocation (land). This example has been used by Niermann [NIE 05] for data ordering tasks, and the author aimed to reveal a diagonal of homogeneous blocks by using a genetic algorithm.

| | A | B | C | D | E | F | G | H | I | J | K | L | M | N | O | P | $x_{i.}$ |
|---|---|---|---|---|---|---|---|---|---|---|---|---|---|---|---|---|---|
| hsco | 0 | 0 | 0 | 0 | 0 | 0 | 0 | 1 | 0 | 0 | 1 | 0 | 0 | 0 | 0 | 0 | 2 |
| agri | 0 | 1 | 1 | 0 | 0 | 0 | 1 | 0 | 0 | 0 | 0 | 1 | 0 | 0 | 1 | 0 | 5 |
| rail | 0 | 0 | 0 | 0 | 0 | 0 | 0 | 1 | 0 | 0 | 1 | 0 | 0 | 0 | 0 | 0 | 2 |
| osco | 1 | 0 | 0 | 0 | 1 | 1 | 0 | 0 | 1 | 1 | 0 | 0 | 1 | 1 | 0 | 1 | 8 |
| vete | 0 | 1 | 1 | 1 | 0 | 0 | 1 | 0 | 0 | 0 | 0 | 1 | 0 | 0 | 1 | 0 | 6 |
| nodo | 1 | 0 | 0 | 0 | 1 | 1 | 0 | 0 | 1 | 1 | 0 | 0 | 1 | 1 | 0 | 1 | 8 |
| nwat | 0 | 0 | 0 | 0 | 0 | 0 | 0 | 0 | 1 | 1 | 0 | 0 | 1 | 1 | 0 | 0 | 4 |
| poli | 0 | 0 | 0 | 0 | 0 | 0 | 0 | 1 | 0 | 0 | 1 | 0 | 0 | 0 | 0 | 0 | 2 |
| land | 0 | 1 | 1 | 1 | 0 | 0 | 1 | 0 | 0 | 0 | 0 | 1 | 0 | 0 | 1 | 0 | 6 |
| $x_{.j}$ | 2 | 3 | 3 | 2 | 2 | 2 | 3 | 3 | 3 | 3 | 3 | 3 | 3 | 3 | 3 | 2 | 43 |

**Table 3.4.** *Townships data matrix*

After having applied LBVEM and LBCEM with different initializations, we retain the best result corresponding to the higher log-likelihood for LBVEM or complete-data log-likelihood for LBCEM. Both algorithms give the same results in terms of co-clustering. The parameters, the partitions, the summary matrix and the reorganized data matrix obtained with LBCEM and the model $[\pi, \rho, \varepsilon_{k\ell}]$ are reported, respectively, in Tables 3.5–3.8.

$$a_{k\ell} = \begin{pmatrix} 0 & 1 & 0 \\ 0 & 0 & 1 \\ 1 & 0 & 0 \end{pmatrix} \quad \varepsilon_{k\ell} = \begin{pmatrix} 0.00 & 0.06 & 0.00 \\ 0.00 & 0.00 & 0.00 \\ 0.17 & 0.00 & 0.00 \end{pmatrix} \quad \alpha_{k\ell} = \begin{pmatrix} 0.00 & 0.94 & 0.00 \\ 0.00 & 0.00 & 1.00 \\ 0.83 & 0.00 & 0.00 \end{pmatrix}$$

**Table 3.5.** *Parameters obtained by* LBVEM

| | |
|---|---|
| 1: agri,vete,land | 1: A, E, F, I, J, M, N, P |
| 2: hsco, rail, poli | 2: B, C, D, G, L, O |
| 3: osco, nodo, nwat | 3: H, K |

**Table 3.6.** *Row (left) and column (right) partitions obtained by* LBVEM

$$\mathbf{x}^{\mathbf{zw}} = \begin{pmatrix} 0 & 17 & 0 \\ 0 & 0 & 6 \\ 20 & 0 & 0 \end{pmatrix}$$

**Table 3.7.** *Summary matrix obtained by* LBVEM

| | AEF IJMNP | BCDGLO | HK |
|---|---|---|---|
| agri | | 11 111 | |
| vete | | 111111 | |
| land | | 111111 | |
| hsco | | | 11 |
| rail | | | 11 |
| poli | | | 11 |
| osco | 11111 111 | | |
| nodo | 11111 111 | | |
| nwat | 11 1 1 | | |

**Table 3.8.** *Reorganized data matrix obtained by* LBVEM

It clearly appears that we can characterize each cluster of townships by a cluster of characteristics: {H, K} by {High School, Railway Station, Police Station}, {B, C, D, G, L, O} by {Agricult Coop, Veterinary, Land Reallocation} and {M, N, J, I, A, P, F, E} by {One Room School, No Doctor, No Water Supply}.

To quickly view the results, a graphical representation of the results can be used (see Figure 3.4).

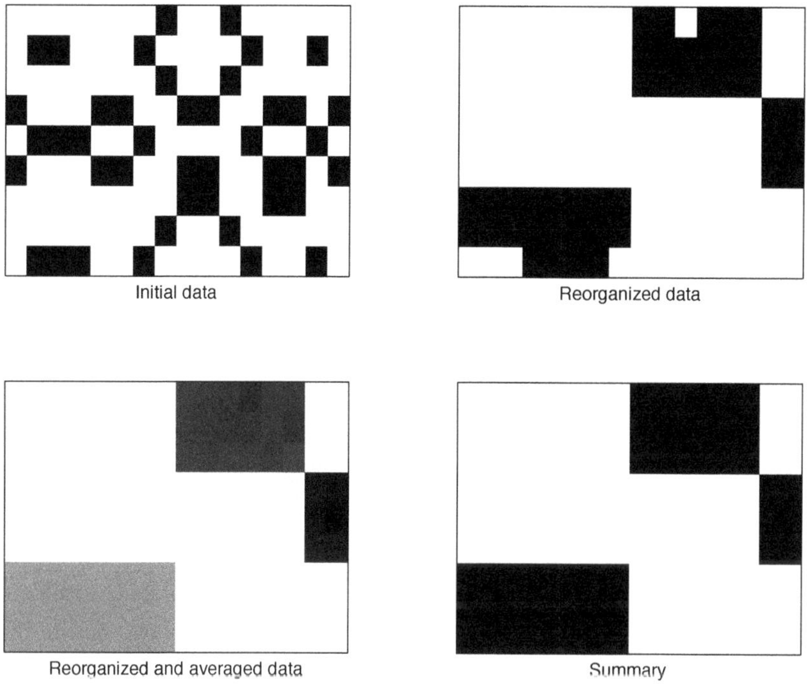


**Figure 3.4.** *Initial and reorganized data according to row and column partitions (top), reorganized averaged data expressing the degree of homogeneity and summary (bottom)*

### 3.8.2. *Mero*

The Mero data set consists of 59 Merovingian belt buckles {bu1,...,bu59} from north-eastern France, ranging from the late sixth Century and the early eighth Century on which the presence or absence of 26 technical criteria of manufacturing, shape and decor {ch1,...,ch26} was observed [LER 80]. The parameters, the row partition, the column partition and the summary matrix obtained with the algorithm LBCEM and the model $[\pi, \rho, \varepsilon_{k\ell}]$ are, respectively, reported in Tables 3.9–3.12 and in Figure 3.5.

$$\alpha_{k\ell} = \begin{pmatrix} 0 & 1 & 0 & 1 & 0 \\ 0 & 1 & 0 & 1 & 0 \\ 0 & 0 & 0 & 1 & 0 \\ 1 & 0 & 0 & 1 & 0 \\ 0 & 0 & 1 & 0 & 0 \end{pmatrix} \quad \varepsilon_{k\ell} = \begin{pmatrix} 0.03 & 0.09 & 0.00 & 0.00 & 0.02 \\ 0.00 & 0.22 & 0.16 & 0.18 & 0.03 \\ 0.00 & 0.35 & 0.06 & 0.03 & 0.31 \\ 0.15 & 0.02 & 0.00 & 0.00 & 0.00 \\ 0.00 & 0.07 & 0.07 & 0.00 & 0.29 \end{pmatrix}$$

$$\alpha_{k\ell} = \begin{pmatrix} 0.03 & 0.91 & 0.00 & 1.00 & 0.02 \\ 0.00 & 0.78 & 0.16 & 0.82 & 0.03 \\ 0.00 & 0.35 & 0.06 & 0.97 & 0.31 \\ 0.85 & 0.02 & 0.00 & 1.00 & 0.00 \\ 0.00 & 0.07 & 0.93 & 0.00 & 0.29 \end{pmatrix}$$

**Table 3.9.** *Parameters obtained by* LBVEM *on the Mero data*

1: bu3,bu9,bu14,bu24,bu25,bu26,bu30,bu32,bu33,bu36,bu46,bu49,bu53
2: bu4,bu20,bu21,bu22,bu31
3: bu7,bu11,bu17,bu27,bu28,bu34,bu35,bu48,bu50,bu51,bu52, bu59
4: bu5,bu15,bu23,bu47,bu54,bu55,bu56,bu57,bu58
5: bu1,bu2,bu6,bu8,bu10,bu12,bu13,bu16,bu18,bu19,bu29,bu37,bu38, bu39,bu40,bu41,bu42,bu43,bu44,bu45

**Table 3.10.** *Row partition obtained by* LBVEM *on the Mero data*

1: ch1,ch6,ch12
2: ch5,ch13,ch17,ch22,ch23
3: ch4,ch8,ch10,ch20,ch21,ch26
4: ch3,ch7,ch9
5: ch2,ch11,ch14,ch15,ch16,ch18,ch19,ch24,ch25

**Table 3.11.** *Column partition obtained by* LBVEM *on the Mero data*

$$\mathbf{x}^{\mathbf{zw}} = \begin{pmatrix} 1 & 58 & 0 & 39 & 2 \\ 0 & 20 & 7 & 11 & 1 \\ 0 & 22 & 4 & 35 & 33 \\ 23 & 1 & 0 & 27 & 0 \\ 0 & 7 & 112 & 0 & 52 \end{pmatrix}$$

**Table 3.12.** *Summary matrix obtained by* LBVEM *on the Mero data*

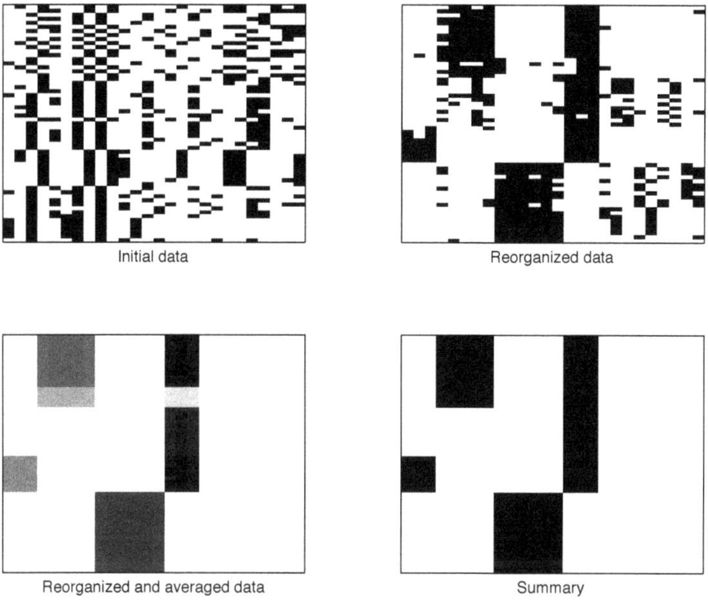


**Figure 3.5.** *Initial and reorganized data according to row and column partitions (top), reorganized averaged data expressing the degree of homogeneity and summary (bottom)*

First, we see the initial data and their reorganization according to the row and column clusters highlighting a structure into homogeneous blocks. Second, we note that this homogeneity can be evaluated with the reorganized averaged data and their summary. Furthermore, we complete the co-clustering results by the projections of row and column clusters obtained with the correspondence analysis (CA) [BEN 73b] in Figures 3.6 and 3.7. Observing the row and column clusters in both planes, we can interpret the associations between the clusters; however, we only have part of the information. Note that the co-clustering directly offers this interpretation.

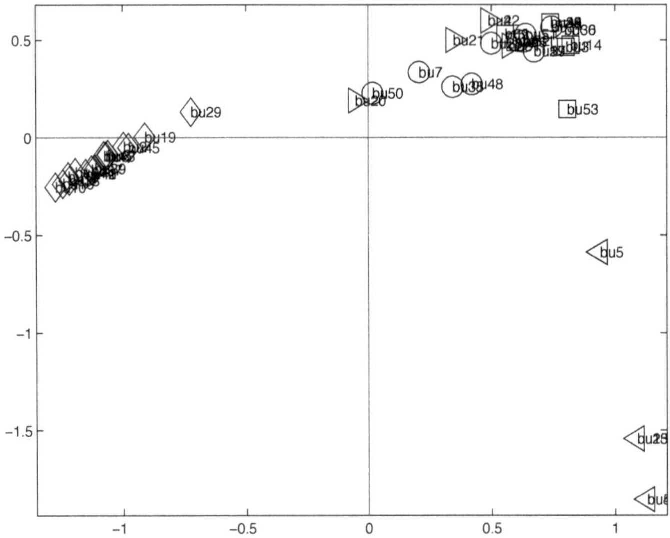


**Figure 3.6.** *Projection of the row clusters into the factorial plane spawned by the first and second axes*

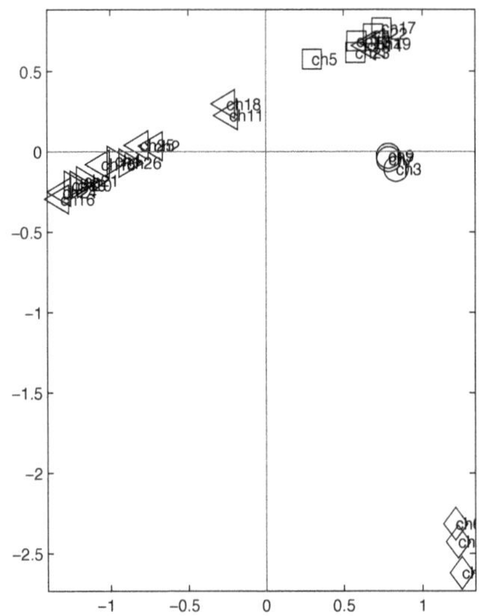


**Figure 3.7.** *Projection of the column clusters into the factorial plane spawned by the first and second axes*

## 3.9. Conclusion

Placing the problem of co-clustering within the maximum likelihood and classification maximum likelihood approaches, we have described two algorithms LBVEM and LBCEM. In terms of co-clustering and estimation, LBVEM appears more efficient than LBCEM. However, as we have emphasized, LBCEM has the advantage of the speed of convergence. Besides, note that although LBVEM does not maximize the likelihood, as in the classical mixture model situation, but only maximizes a fuzzy criterion, the variational approximation seems good. However, and despite the quality of the results obtained by the approximation proposed and commonly used in physics as a means to avoid the heavy computation involved in the E step of EM (see, for instance, [HOF 99a]), the link between the fuzzy criterion and the true log-likelihood, and not its approximation, deserves future investigation.

Chapter 4

# Co-Clustering of Contingency Tables

The co-clustering methods have practical importance in a wide variety of applications such as document clustering where the data are often arranged as two-way contingency tables. In this situation (see the Introduction), the sets $I$ and $J$ are two categorical variables. Therefore the rows and columns correspond to different categories of the two variables and the data matrix, which displays the frequency distribution of the variables, is the contingency table. The sets $I$ and $J$ may also be any two sets with a data matrix defining a binary relation on $I \times J$. This chapter presents a coherent framework to understand some existing criteria and algorithms for analyzing contingency tables and to propose new tables. Two approaches of the problem of co-clustering are studied.

– The first approach is based on the minimization of an objective function based on the measures of association. Two algorithms, called CROKI2 and CROINFO, based on chi-squared statistic and mutual information are studied. Note that the proposed formalism can be extended to other measures of association. Furthermore, links between these two algorithms can be established, and finally, we justify the use of correspondence analysis (CA) as a method of visualization.

– The second approach is based on a probabilistic approach. In this situation, two models are considered: the *block model* that gives a probabilistic interpretation of the criteria optimized by CROKI2 and CROINFO and the *latent block model* (LBM) that offers not only a probabilistic interpretation by using Poisson distributions, also but new criteria and algorithms in the family of the expectation-maximization algorithm [DEM 77], referred to hereafter as PLBVEM and PLBCEM. The latter can offer rich perspectives in the selection model context, a problem that we do not deal with here.

This chapter is organized as follows. In sections 4.1 and 4.2, we give the necessary background on measures of association used in this chapter. In section 4.3, we present the co-clustering approach based on these association measures. Section 4.4 deals with model-based co-clustering: a block model and an LBM recently developed in [GOV 10]. Fuzzy and hard co-clustering algorithms are described in detail and some connections between them are established. In section 4.5, we focus on the comparison and illustrations of our approach.

## 4.1. Measures of association

The contingency table characterizes the dependency links between two sets $I$ and $J$, and measuring the strength of this association is a long tradition in statistics, going back at least to the work of Pearson [PEA 00]. In this chapter, two measures of association defined in the following sections will be used.

### 4.1.1. *Phi-squared coefficient*

Many measures of association arise from the standard chi-squared statistic upon which a test of independence is usually based

$$\chi^2(\mathbf{x}) = \sum_{i,j} \frac{(x_{ij} - x_{i.}x_{.j}/N)^2}{x_{i.}x_{.j}/N}$$

where $x_{i.}$, $x_{.j}$ and $N$ are notations defined in Introduction.

The $\chi^2$ measure usually provides statistical evidence of a significant association, or dependency, between rows and columns in the table. The phi-squared Pearson's coefficient of mean squared contingency is based on this $\chi^2$ statistic as follows:

$$\Phi^2(P_{IJ}) = \frac{\chi^2(\mathbf{x})}{N} = \sum_{i,j} \frac{(p_{ij} - p_{i.}p_{.j})^2}{p_{i.}p_{.j}}$$

$$= \sum_{i,j} \frac{p_{ij}^2}{p_{i.}p_{.j}} - 1 = \sum_{i,j} p_{i.}p_{.j}(\frac{p_{ij}}{p_{i.}p_{.j}} - 1)^2.$$

Considering $\mathcal{N}$ the cloud of $I \times J$, the set of $n \times d$ points belonging to $\mathbb{R}$ weighted by $p_{i.}p_{.j}$, the center of gravity of this cloud is therefore equal to 1. The last formulation shows that the discrepancy of the independence model is taken into account by considering $x_{ij} = \frac{p_{ij}}{p_{i.}p_{.j}} - 1$ as a general term of x. More generally, the phi-squared coefficient can be seen as an estimation of the deviation between both the probabilities $\xi_{i.}\xi_{.j}$, which we would have if the two categorical random variables were independent, and the probabilities $\xi_{ij}$. The phi-squared coefficient corrects chi-squared sensitivity to sample size by dividing the chi-squared statistic by the number of cases. It can be shown that this coefficient ranges between 0, indicative of complete independence, and

$\min(n, d) - 1$, indicative of perfect association. Since its range depends on the dimensions of the table and to overcome this difficulty, variations of this measure have been proposed as the Pearson's contingency coefficient $\sqrt{\frac{\Phi^2}{1+\Phi^2}}$, the Tschuprow's contingency coefficient $t = \frac{\Phi^2}{\sqrt{(n-1)(d-1)}}$ or the Cramer's contingency coefficient $v = \sqrt{\frac{\Phi^2}{min(n,d)-1}}$.

Note also that this chi-squared statistic measure of association is the "base" of the CA. The CA method is an exploratory multivariate technique that converts a contingency table into a particular type of graphical display in which the rows and the columns of the matrix are depicted as points [BEN 73b, GRE 07]. It aims to study the relationship between two categorical variables, precisely by describing the variability of the row and column profiles. It can be used on any two-way table, sparse or not, as the case may be. It projects the rows and columns of a data matrix onto points within a graph in a Euclidean space. The graphs or maps that are the factorial planes spawned by the different axes are then used to gain some understanding of the data and to extract information from them. Typically, this method of visualization proved its usefulness by giving a representation of the rows and columns. Hence, CA and co-clustering in this context can be more beneficial for contingency tables. For instance, consider that our data set in Table 4.1 is a contingency table resulting from the two sets of rows and columns $\{r1, \ldots, r6\}$ and $\{c1, \ldots, c5\}$. By applying CA onto this table, we obtain a good representation shown in Figure 4.1; the percentage of the total inertia explained by the factorial plane spawned by the first and second axes is equal to 98.87%. Note that we can observe some clusters such as $\{r_1, r_2\}$, $\{r3, r4\}$, $\{r5, r6\}$, $\{c1, c2, c3\}$ and $\{c4, c5\}$.

| | 1 | 2 | 3 | 4 | 5 | |
|---|---|---|---|---|---|---|
| 1 | 5 | 4 | 6 | 1 | 0 | 16 |
| 2 | 6 | 5 | 4 | 0 | 1 | 16 |
| 3 | 1 | 0 | 1 | 7 | 5 | 14 |
| 4 | 1 | 1 | 0 | 6 | 5 | 13 |
| 5 | 4 | 5 | 3 | 4 | 5 | 21 |
| 6 | 5 | 4 | 4 | 3 | 4 | 20 |
| | 22 | 19 | 18 | 21 | 20 | 100 |

| | 1 | 2 | 3 | 4 | 5 | |
|---|---|---|---|---|---|---|
| 1 | 0.05 | 0.04 | 0.06 | 0.01 | 0.00 | 0.16 |
| 2 | 0.06 | 0.05 | 0.04 | 0.00 | 0.01 | 0.16 |
| 3 | 0.01 | 0.00 | 0.01 | 0.07 | 0.05 | 0.14 |
| 4 | 0.01 | 0.01 | 0.00 | 0.06 | 0.05 | 0.13 |
| 5 | 0.04 | 0.05 | 0.03 | 0.04 | 0.05 | 0.21 |
| 6 | 0.05 | 0.04 | 0.04 | 0.03 | 0.04 | 0.20 |
| | 0.22 | 0.19 | 0.18 | 0.21 | 0.20 | 1.00 |

**Table 4.1.** *Example of contingency table and associated joint distribution*

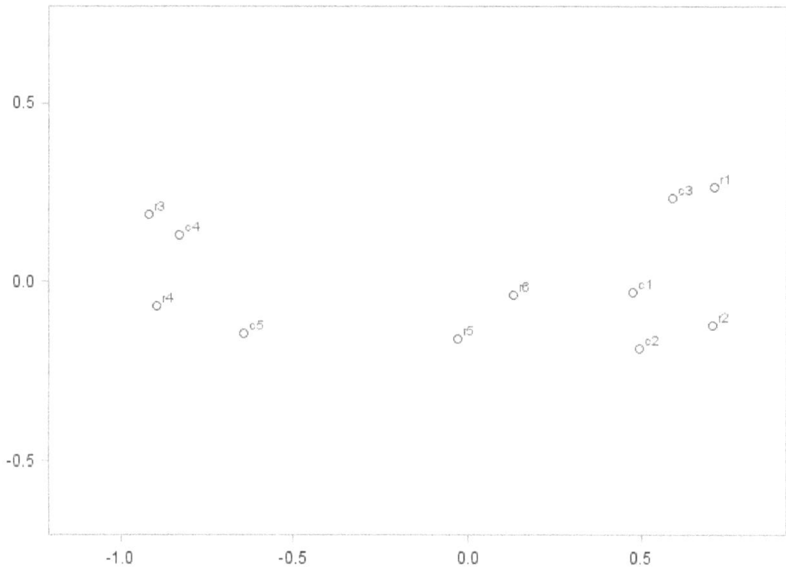


**Figure 4.1.** *CA: projection of the rows and columns into the factorial plane spawned by the first and second axes of CA*

### 4.1.2. *Mutual information*

To characterize the association between two categorical variables, the likelihood-ratio statistic, also called likelihood ratio chi-squared,

$$G^2(\mathbf{x}) = 2\sum_{i,j} x_{ij} \log \frac{x_{ij}}{x_{i.}x_{.j}/N}$$

is an alternative to the chi-squared statistic. It statistically compares the maximum likelihood of an unrestricted model with parameters $\xi_{ij}$ with a restricted model with parameters $\xi_{i.}\xi_{.j}$. In this setting, the unrestricted model consists of the observed frequencies in the data and the restricted model consists of the expected frequencies under the null hypothesis of no association. The normalization of this measure yields the mutual information measure

$$\mathcal{I}(P_{IJ}) = \frac{G^2(\mathbf{x})}{2N} = \sum_{i,j} p_{ij} \log \frac{p_{ij}}{p_{i.}p_{.j}},$$

an information-theoretic notion usually defined by

$$\mathcal{I}(P_{IJ}) = \mathrm{H}(P_I) + \mathrm{H}(P_J) - \mathrm{H}(P_{IJ})$$

where $\mathrm{H}(P_I) = -\sum_i p_{i.} \log p_{i.}$ and $\mathrm{H}(P_J) = -\sum_j p_{.j} \log p_{.j}$ are the marginal entropies and $\mathrm{H}(P_{IJ}) = -\sum_{i,j} p_{ij} \log p_{ij}$ is the joint entropy of $I$ and $J$.

Note that there is a link between phi-squared coefficient and mutual information described in the following proposition:

PROPOSITION 4.1.– $\mathcal{I}(P_{IJ}) = \frac{1}{2}\Phi^2(P_{IJ}) + o(\sum_{ij}(p_{ij} - p_{i.}p_{.j})^2)$.

PROOF.– Denoting $e_{ij} = p_{ij} - p_{i.}p_{.j}$ for all $i, j$ and using the following equation $\log(1+x) = x - x^2/2 + o(x^2)$, it can be written as

$$\log(1 + \frac{e_{ij}}{p_{i.}p_{.}}) = \frac{e_{ij}}{p_{i.}p_{.j}} - \frac{1}{2}\left(\frac{e_{ij}}{p_{i.}p_{.j}}\right)^2 + o(e_{ij}^2),$$

$$(p_{ij} + e_{ij}) \log(1 + \frac{e_{ij}}{p_{i.}p_{.}}) = e_{ij} - \frac{1}{2}\frac{e_{ij}^2}{p_{i.}p_{.j}} + \frac{e_{ij}^2}{p_{i.}p_{.j}} - \frac{1}{2}\frac{e_{ij}^3}{p_{i.}^2 p_{.j}^2} + o(e_{ij}^2)$$

$$= e_{ij} + \frac{1}{2}\frac{e_{ij}^2}{p_{i.}p_{.j}} + o(e_{ij}^2)$$

and therefore

$$\sum_{i,j}(p_{ij}+e_{ij})\log(1+\frac{e_{ij}}{p_{i.}p_{.}}) = \underbrace{\sum_{i,j} e_{ij}}_{=0} + \frac{1}{2}\sum_{i,j}\frac{e_{ij}^2}{p_{i.}p_{.j}} + o(\sum_{i,j} e_{ij}^2),$$

which shows the proposition. ■

This link justifies the use of CA, which is based on $\Phi^2$, as an appropriate visualization method for the rows and columns of contingency tables used in our examples.

## 4.2. Contingency table associated with a couple of partitions

In this section, we study the effect of simultaneously aggregating the rows and the columns of a contingency table x according to a couple of partitions of the sets $I$ and $J$. If z and w are partitions in $g$ clusters and $m$ clusters of the set $I$ of the rows and the set $J$ of columns of the contingency table x, a new two-way contingency table $\mathbf{x}^{\mathbf{zw}} = (x_{k\ell}^{\mathbf{zw}})$ can be associated with two categorical random variables that take values in sets $K = \{1, \ldots, g\}$ and $L = \{1, \ldots, m\}$ by merging the rows and columns according to the partitions z and w

$$x_{k\ell}^{\mathbf{zw}} = \sum_{i,j} z_{ik} w_{j\ell} x_{ij} \qquad \forall k \in K \quad \text{and} \quad \forall \ell \in L.$$

### 4.2.1. *Associated distributions*

The first distribution that can be associated with z and w is the distribution $P_{KL}^{\mathbf{zw}} = (p_{k\ell}^{\mathbf{zw}})$ defined on $K \times L$ by

$$p_{k\ell}^{\mathbf{zw}} = \frac{x_{k\ell}^{\mathbf{zw}}}{N} = \sum_{i,j} z_{ik} w_{j\ell}\, p_{ij} \qquad \forall (k,\ell) \in K \times L.$$

The following equation

$$\sum_{k,\ell} p^{\mathbf{zw}}_{k\ell} = \sum_{i,j,k,\ell} z_{ik} w_{j\ell} p_{ij} = \sum_{i,j} p_{ij} \underbrace{\sum_{k,\ell} z_{ik} w_{j\ell}}_{=1} = 1$$

shows that $P^{\mathbf{zw}}_{KL}$ is a distribution. Moreover, it can be noted that, as

$$\sum_{\ell} p^{\mathbf{zw}}_{k\ell} = \sum_{i,j,\ell} z_{ik} w_{j\ell}\, p_{ij} = \sum_{i} (z_{ik} \sum_{j} (p_{jj} \underbrace{\sum_{\ell} w_{j\ell}}_{=1})) = \sum_{i} z_{ik} p_{i.},$$

the row margins of this new distribution do not depend on the partition w and will be denoted as $p^{\mathbf{z}}_{k.}$. Similarly, the column margins $\sum_k p^{\mathbf{zw}}_{k\ell}$ are equal to $\sum_j w_{j\ell} p_{.j}$ and will be denoted as $p^{\mathbf{w}}_{.\ell}$.

The second distribution that can be associated with z and w is the distribution $Q^{\mathbf{zw}}_{IJ} = (q^{\mathbf{zw}}_{ij})$ defined on $I \times J$ by

$$q^{\mathbf{zw}}_{ij} = p_{i.} p_{.j} \sum_{k,\ell} z_{ik} w_{j\ell} \frac{p^{\mathbf{zw}}_{k\ell}}{p^{\mathbf{z}}_{k.} p^{\mathbf{w}}_{.\ell}} \qquad \forall (i,j) \in I \times J.$$

The following equation

$$\sum_{i,j} q^{\mathbf{zw}}_{ij} = \sum_{i,j} p_{i.} p_{.j} \sum_{k,\ell} z_{ik} w_{j\ell} \frac{p^{\mathbf{zw}}_{k\ell}}{p^{\mathbf{z}}_{k.} p^{\mathbf{w}}_{.\ell}} = \sum_{k,\ell} \frac{p^{\mathbf{zw}}_{k\ell}}{p^{\mathbf{z}}_{k.} p^{\mathbf{w}}_{.\ell}} \underbrace{\sum_{i,j} p_{i.} p_{.j} z_{ik} w_{j\ell}}_{=p^{\mathbf{z}}_{k.} p^{\mathbf{w}}_{.\ell}}$$

$$= \sum_{k,\ell} p^{\mathbf{zw}}_{k\ell} = 1,$$

shows that $Q_{IJ}^{\mathbf{zw}}$ is a distribution. Moreover, as

$$q_{i.}^{\mathbf{z}} = \sum_j q_{ij}^{\mathbf{zw}} = \sum_j p_{i.}p_{.j} \sum_{k,\ell} z_{ik} w_{j\ell} \frac{p_{k\ell}^{\mathbf{zw}}}{p_{k.}^{\mathbf{z}} p_{.\ell}^{\mathbf{w}}}$$

$$= p_{i.} \sum_k \frac{z_{ik}}{p_{k.}^{\mathbf{z}}} \underbrace{\sum_\ell \frac{\overbrace{\sum_j (w_{j\ell} p_{.j})}^{p_{.\ell}^{\mathbf{w}}} p_{k\ell}^{\mathbf{zw}}}{p_{.\ell}^{\mathbf{w}}}}_{=p_{k.}^{\mathbf{z}}} = p_{i.}$$

and, symmetrically, $q_{.j}^{\mathbf{w}} = p_{.j}$, this distribution has the same margins as the initial distribution $P_{IJ}$.

Distributions $Q_{IJ}^{\mathbf{zw}}$ and $P_{KL}^{\mathbf{zw}}$ can be illustrated using the example described in Table 4.1. For instance, the aggregation of the rows and columns of the data according to the partitions $\mathbf{z} = (1, 1, 2, 2, 3, 3)$ and $\mathbf{w} = (1, 1, 1, 2, 2)$ leads to the contingency table $\mathrm{x}^{\mathbf{zw}}$ and the distribution $P_{KL}^{\mathbf{zw}}$ presented in Table 4.2.

| | 1 | 2 | |
|---|---|---|---|
| 1 | 30.0 | 2.00 | 32.0 |
| 2 | 4.00 | 23.0 | 27.0 |
| 3 | 25.0 | 16.0 | 41.0 |
| | 59.0 | 41.0 | 100 |

| | 1 | 2 | |
|---|---|---|---|
| 1 | 0.30 | 0.02 | 0.32 |
| 2 | 0.04 | 0.23 | 0.27 |
| 3 | 0.25 | 0.16 | 0.41 |
| | 0.59 | 0.41 | 1.00 |

**Table 4.2.** *Aggregated contingency table* $\mathrm{x}^{\mathbf{zw}}$ *(left) and associated distribution* $P_{KL}^{\mathbf{zw}}$ *(right)*

Table 4.3 gives the original distribution $P_{IJ}$ and the distribution $Q_{IJ}^{\mathbf{zw}}$ obtained after aggregating the rows and columns. It can be seen that, in this example, these two distributions are similar. Looking at the factorial plan in Figure 4.1, associations between row clusters and column clusters can be observed. We can then make sense of the row clusters and column clusters simultaneously: $\{r_1, r_2\}$

characterized by high values in $\{c1, c2, c3\}$ and small values in $\{c4, c5\}$, $\{r3, r4\}$ characterized by high values in $\{c4, c5\}$ and small values in $\{c1, c2, c3\}$ and $\{r5, r6\}$ characterized by high values in $\{c1, c2, c3\}$ and moderate values in $\{c4, c5\}$.

| | 1 | 2 | 3 | 4 | 5 | |
|---|---|---|---|---|---|---|
| 1 | 0.050 | 0.040 | 0.060 | 0.010 | 0.000 | 0.160 |
| 2 | 0.060 | 0.050 | 0.040 | 0.000 | 0.010 | 0.160 |
| 3 | 0.010 | 0.000 | 0.010 | 0.070 | 0.050 | 0.140 |
| 4 | 0.010 | 0.010 | 0.000 | 0.060 | 0.050 | 0.130 |
| 5 | 0.040 | 0.050 | 0.030 | 0.040 | 0.050 | 0.210 |
| 6 | 0.050 | 0.040 | 0.040 | 0.030 | 0.040 | 0.200 |
| | 0.220 | 0.190 | 0.180 | 0.210 | 0.200 | 1.000 |

| | 1 | 2 | 3 | 4 | 5 | |
|---|---|---|---|---|---|---|
| 1 | 0.056 | 0.048 | 0.046 | 0.005 | 0.005 | 0.160 |
| 2 | 0.056 | 0.048 | 0.046 | 0.005 | 0.005 | 0.160 |
| 3 | 0.008 | 0.007 | 0.006 | 0.061 | 0.058 | 0.140 |
| 4 | 0.007 | 0.006 | 0.006 | 0.057 | 0.054 | 0.130 |
| 5 | 0.048 | 0.041 | 0.039 | 0.042 | 0.040 | 0.210 |
| 6 | 0.045 | 0.039 | 0.037 | 0.040 | 0.038 | 0.200 |
| | 0.220 | 0.190 | 0.180 | 0.210 | 0.200 | 1.000 |

**Table 4.3.** *Distributions $P_{IJ}$ (left) and $Q_{IJ}^{\mathbf{zw}}$ (right)*

### 4.2.2. *Associated measures of association*

Using the two measures, phi-squared coefficient and mutual information, applied to the two distributions previously defined, we obtain the following measures

$$\Phi^2(P_{KL}^{\mathbf{zw}}) = \sum_{k,\ell} \frac{(p_{k\ell}^{\mathbf{zw}} - p_{k.}^{\mathbf{z}} p_{.\ell}^{\mathbf{w}})^2}{p_{k.}^{\mathbf{z}} p_{.\ell}^{\mathbf{w}}}, \quad \Phi^2(Q_{IJ}^{\mathbf{zw}}) = \sum_{i,j} \frac{(q_{ij}^{\mathbf{zw}} - p_{i.}p_{.j})^2}{p_{i.}p_{.j}}$$

$$\mathcal{I}(P_{KL}^{\mathbf{zw}}) = \sum_{k,\ell} p_{k\ell}^{\mathbf{zw}} \log \frac{p_{k\ell}^{\mathbf{zw}}}{p_{k.}^{\mathbf{z}} p_{.\ell}^{\mathbf{w}}} \quad \text{and} \quad \mathcal{I}(Q_{IJ}^{\mathbf{zw}}) = \sum_{i,j} q_{ij}^{\mathbf{zw}} \log \frac{q_{ij}^{\mathbf{zw}}}{p_{i.}p_{.j}}.$$

For the rest of this chapter, it is interesting to establish the relationship among the measures of association of the three distributions $P_{IJ}$, $P_{KL}^{\mathbf{zw}}$ and $Q_{IJ}^{\mathbf{zw}}$. These properties, which are similar for both types of measure of association, are grouped into the following two propositions.

PROPOSITION 4.2.–

$$\Phi^2(P_{KL}^{\mathbf{zw}}) = \Phi^2(Q_{IJ}^{\mathbf{zw}}),$$

$$\Phi^2(P_{IJ}) - \Phi^2(Q_{IJ}^{\mathbf{zw}}) = \Phi^2(P_{IJ}) - \Phi^2(P_{KL}^{\mathbf{zw}}) = D_{\Phi^2}(P_{IJ}||Q_{IJ}^{\mathbf{zw}})$$

where

$D_{\Phi^2}(P_{IJ}||Q_{IJ}^{\mathbf{zw}}) = \sum_{i,j} \frac{(p_{ij}-q_{ij}^{\mathbf{zw}})^2}{p_{i.}p_{.j}} = \sum_{i,j} p_{ij}\left(\frac{p_{ij}}{p_{i.}p_{.j}} - \frac{q_{ij}^{\mathbf{zw}}}{p_{i.}p_{.j}}\right)$ **can be viewed as a $\Phi^2$ distance between the distributions $P_{IJ}$ and $Q_{IJ}^{\mathbf{zw}}$**, $\Phi^2(Q_{IJ}^{\mathbf{zw}}) \leq \Phi^2(P_{IJ})$ **or** $\Phi^2(P_{KL}^{\mathbf{zw}}) \leq \Phi^2(P_{IJ})$.

LEMMA 4.1.–

$$\sum_{k,\ell} \frac{(p_{k\ell}^{\mathbf{zw}})^2}{p_{k.}^{\mathbf{z}} p_{.\ell}^{\mathbf{w}}} = \sum_{i,j} \frac{p_{ij} q_{ij}^{\mathbf{zw}}}{p_{i.}p_{.j}} = \sum_{i,j} \frac{(q_{ij}^{\mathbf{zw}})^2}{p_{i.}p_{.j}}$$

PROOF.–

$$\sum_{i,j} \frac{p_{ij} q_{ij}^{\mathbf{zw}}}{p_{i.}p_{.j}} = \sum_{i,j} p_{ij} \frac{p_{i.}p_{.j} \sum_{k,\ell} z_{ik} w_{j\ell} \frac{p_{k\ell}^{\mathbf{zw}}}{p_{k.}^{\mathbf{z}} p_{.\ell}^{\mathbf{w}}}}{p_{i.}p_{.j}}$$

$$= \sum_{k,\ell} \frac{p_{k\ell}^{\mathbf{zw}}}{p_{k.}^{\mathbf{z}} p_{.\ell}^{\mathbf{w}}} \sum_{i,j} z_{ik} w_{j\ell} p_{ij} = \sum_{k,\ell} \frac{(p_{k\ell}^{\mathbf{zw}})^2}{p_{k.}^{\mathbf{z}} p_{.\ell}^{\mathbf{w}}}$$

$$\sum_{i,j} \frac{(q_{ij}^{\mathbf{zw}})^2}{p_{i.}p_{.j}} = \sum_{i,j} \frac{\left(p_{i.}p_{.j} \sum_{k,\ell} z_{ik} w_{j\ell} \frac{p_{k\ell}^{\mathbf{zw}}}{p_{k.}^{\mathbf{z}} p_{.\ell}^{\mathbf{w}}}\right)^2}{p_{i.}p_{.j}}$$

$$= \sum_{i,j} p_{i.}p_{.j} \sum_{k,\ell} z_{ik} w_{j\ell} \frac{(p_{k\ell}^{\mathbf{zw}})^2}{(p_{k.}^{\mathbf{z}} p_{.\ell}^{\mathbf{w}})^2} = \sum_{k,\ell} \frac{(p_{k\ell}^{\mathbf{zw}})^2}{p_{k.}^{\mathbf{z}} p_{.\ell}^{\mathbf{w}}}.$$

■

PROOF.– The first equation can easily be deduced from lemma 4.1. We have

$$\Phi^2(P_{IJ}) - \Phi^2(Q_{IJ}^{\mathbf{zw}}) = \left(\sum_{i,j} \frac{(p_{ij})^2}{p_{i.}p_{.j}} - 1\right) - \left(\sum_{i,j} \frac{(q_{ij}^{\mathbf{zw}})^2}{q_{i.}^{\mathbf{z}} q_{.j}^{\mathbf{w}}} - 1\right)$$

$$= \sum_{i,j} \frac{p_{ij}^2}{p_{i.}p_{.j}} - \sum_{i,j} \frac{(q_{ij}^{\mathbf{zw}})^2}{p_{i.}p_{.j}}$$

and using lemma [4.1], this expression can be written as

$$\sum_{i,j} \frac{p_{ij}^2}{p_{i.}p_{.j}} - \sum_{i,j} \frac{p_{ij}q_{ij}^{\mathbf{zw}}}{p_{i.}p_{.j}} = \sum_{i,j} p_{ij} \left( \frac{p_{ij}}{p_{i.}p_{.j}} - \frac{q_{ij}^{\mathbf{zw}}}{p_{i.}p_{.j}} \right)$$

and

$$\sum_{i,j} \frac{p_{ij}^2}{p_{i.}p_{.j}} - 2\sum_{i,j} \frac{p_{ij}q_{ij}^{\mathbf{zw}}}{p_{i.}p_{.j}} + \sum_{i,j} \frac{(q_{ij}^{\mathbf{zw}})^2}{p_{i.}p_{.j}} = \sum_{i,j} \frac{(p_{ij} - q_{ij}^{\mathbf{zw}})^2}{p_{i.}p_{.j}}$$

that shows the second equation. The third equation can easily be deduced from the previous equation. ■

PROPOSITION 4.3.–

$$\mathcal{I}(P_{KL}^{\mathbf{zw}}) = \mathcal{I}(Q_{IJ}^{\mathbf{zw}})$$

$$\mathcal{I}(P_{IJ}) - \mathcal{I}(Q_{IJ}^{\mathbf{zw}}) = \mathcal{I}(P_{IJ}) - \mathcal{I}(P_{KL}^{\mathbf{zw}}) = \mathrm{KL}(P_{IJ}||Q_{IJ}^{\mathbf{zw}})$$

where $\mathrm{KL}(P_{IJ}||Q_{IJ}^{\mathbf{zw}}) = \sum_{i,j} p_{ij} \log \frac{p_{ij}}{q_{ij}^{\mathbf{zw}}}$ is the Kullback–Leibler distance between the two distributions $P_{IJ}$ and $Q_{IJ}^{\mathbf{zw}}$),

$$\mathcal{I}(Q_{IJ}^{\mathbf{zw}}) \leq \mathcal{I}(P_{IJ}) \quad \text{or} \quad \mathcal{I}(P_{KL}^{\mathbf{zw}}) \leq \mathcal{I}(P_{IJ}).$$

PROOF.–

$$\begin{aligned}
\mathcal{I}(Q_{IJ}^{\mathbf{zw}}) &= \sum_{i,j} q_{ij}^{\mathbf{zw}} \log \frac{q_{ij}^{\mathbf{zw}}}{p_{i.}p_{.j}} \\
&= \sum_{i,j} \left( p_{i.}p_{.j} \left( \sum_{k,\ell} z_{ik}w_{j\ell} \frac{p_{k\ell}^{\mathbf{zw}}}{p_{k.}^{\mathbf{z}}p_{.\ell}^{\mathbf{w}}} \right) \log \left( \sum_{k,\ell} z_{ik}w_{j\ell} \frac{p_{k\ell}^{\mathbf{zw}}}{p_{k.}^{\mathbf{z}}p_{.\ell}^{\mathbf{w}}} \right) \right) \\
&= \sum_{i,j} \left( p_{i.}p_{.j} \left( \sum_{k,\ell} z_{ik}w_{j\ell} \frac{p_{k\ell}^{\mathbf{zw}}}{p_{k.}^{\mathbf{z}}p_{.\ell}^{\mathbf{w}}} \log \frac{p_{k\ell}^{\mathbf{zw}}}{p_{k.}^{\mathbf{z}}p_{.\ell}^{\mathbf{w}}} \right) \right) \\
&= \sum_{k,\ell} p_{k\ell}^{\mathbf{zw}} \log \frac{p_{k\ell}^{\mathbf{zw}}}{p_{k.}^{\mathbf{z}}p_{.\ell}^{\mathbf{w}}} = \mathcal{I}(P_{KL}^{\mathbf{zw}})
\end{aligned}$$

$$
\begin{aligned}
\mathcal{I}(P_{IJ}) - \mathcal{I}(Q_{IJ}^{\mathbf{zw}}) = \mathcal{I}(P_{IJ}) - \mathcal{I}(P_{KL}^{\mathbf{zw}}) &= \sum_{i,j} p_{ij} \log \frac{p_{ij}}{p_{i.}p_{.j}} - \sum_{k,\ell} p_{k\ell}^{\mathbf{zw}} \log \frac{p_{k\ell}^{\mathbf{zw}}}{p_{k.}^{\mathbf{z}} p_{.\ell}^{\mathbf{w}}} \\
&= \sum_{i,j,k,\ell} z_{ik} w_{j\ell} p_{ij} \log \frac{p_{ij}}{p_{i.}p_{.j}} - \sum_{i,j,k,\ell} (z_{ik} w_{j\ell} p_{ij}) \log \frac{p_{k\ell}^{\mathbf{zw}}}{p_{k.}^{\mathbf{z}} p_{.\ell}^{\mathbf{w}}} \\
&= \sum_{i,j,k,\ell} z_{ik} w_{j\ell} p_{ij} \log \frac{p_{ij}}{p_{i.}p_{.j}} \frac{p_{k.}^{\mathbf{z}} p_{.\ell}^{\mathbf{w}}}{p_{k\ell}^{\mathbf{zw}}} \\
&= \sum_{i,j,k,\ell} z_{ik} w_{j\ell} p_{ij} \log \frac{p_{ij}}{p_{i.}p_{.j} \frac{p_{k\ell}^{\mathbf{zw}}}{p_{k.}^{\mathbf{z}} p_{.\ell}^{\mathbf{w}}}} \\
&= \sum_{i,j} p_{ij} \log \frac{p_{ij}}{p_{i.}p_{.j} \sum_{k,\ell} z_{ik} w_{j\ell} \frac{p_{k\ell}^{\mathbf{zw}}}{p_{k.}^{\mathbf{z}} p_{.\ell}^{\mathbf{w}}}} \\
&= \sum_{i,j} p_{ij} \log \frac{p_{ij}}{q_{ij}} = \mathrm{KL}(P_{IJ} || Q_{IJ}^{\mathbf{zw}}).
\end{aligned}
$$

The third equation can easily be deduced from the previous equation. ■

## 4.3. Co-clustering of contingency table

### 4.3.1. *Two equivalent approaches*

Propositions 4.2 and 4.3 show that the properties of the co-clustering defined by the couple of partitions z and w can equally be expressed in terms of the distribution $P_{KL}^{\mathbf{zw}}$ or distribution $Q_{IJ}^{\mathbf{zw}}$. Two approaches of co-clustering of contingency table can be then considered:

– Co-clustering obtained by reducing the size of the contingency table: looking for a good co-clustering can be seen as a way to obtain a good summary of the data. This aim can be reached by collapsing the rows and columns according to row and column partitions to obtain a reduced contingency $\mathbf{x}^{\mathbf{zw}}$ table that preserves a relationship of interest. Using the two measures of association previously defined, the objective of co-clustering can be based on maximizing

$$\Phi^2(P_{KL}^{\mathbf{zw}}) \quad \text{or} \quad \mathcal{I}(P_{KL}^{\mathbf{zw}}),$$

or, equally, on minimizing

$$\Phi^2(P_{IJ}) - \Phi^2(P_{KL}^{\mathbf{zw}}) \quad \text{or} \quad \mathcal{I}(P_{IJ}) - \mathcal{I}(P_{KL}^{\mathbf{zw}}). \qquad [4.1]$$

– Co-clustering obtained by approximating the initial distribution: the co-clustering problem can be viewed as an approximation of the distribution $P_{IJ}$ by a distribution according to co-clustering termed $Q_{IJ}^{\mathbf{zw}}$ by minimizing the difference between the measures of information of the initial distribution and of the new distribution

$$\Phi^2(P_{IJ}) - \Phi^2(Q_{IJ}^{\mathbf{zw}}) \quad \text{or} \quad \mathcal{I}(P_{IJ}) - \mathcal{I}(Q_{IJ}^{\mathbf{zw}}). \qquad [4.2]$$

Finally, in both situations, the problem is finding the partitions z and w minimizing the criterion for the $\Phi^2$ measure of association

$$\begin{aligned} \mathrm{W}_{\Phi^2}(\mathbf{z},\mathbf{w}) &= D_{\Phi^2}(P_{IJ}||P_{KL}^{\mathbf{zw}}) = \Phi^2(P_{IJ}) - \Phi^2(P_{KL}^{\mathbf{zw}}) \\ &= \Phi^2(P_{IJ}) - \Phi^2(Q_{IJ}^{\mathbf{zw}}) \end{aligned}$$

or minimizing the criterion for the $\mathcal{I}$ measure of association

$$\begin{aligned} \mathrm{W}_{\mathcal{I}}(\mathbf{z},\mathbf{w}) &= \mathrm{KL}(P_{IJ}||P_{KL}^{\mathbf{zw}}) = \mathcal{I}(P_{IJ}) - \mathcal{I}(P_{KL}^{\mathbf{zw}}) \\ &= \mathcal{I}(P_{IJ}) - \mathcal{I}(Q_{IJ}^{\mathbf{zw}}). \end{aligned}$$

It can be shown that there is no loss of information ($\mathrm{W}_{\Phi^2}(\mathbf{z},\mathbf{w}) = \mathrm{W}_{\mathcal{I}}(\mathbf{z},\mathbf{w}) = 0$) if the merged rows and columns have the same profile, that is if $(\frac{p_{i1}}{p_{i.}},\ldots,\frac{p_{id}}{p_{i.}})$ is constant for all $i$ belonging to a same row cluster and if $(\frac{p_{1j}}{p_{.j}},\ldots,\frac{p_{nj}}{p_{.j}})$ is constant for all $j$ belonging to a same column cluster. This property of contingency tables, which is a fundamental property of CA [BEN 73b, GRE 88a] is the so-called principle of *distributional equivalence*. In this situation, it can be noted that the initial distribution $P_{IJ}$ and the final distribution $Q_{IJ}^{\mathbf{zw}}$ are the same.

Next, we focus on two algorithms of co-clustering based on both equivalent approaches. The first algorithm consists of maximizing the expression [4.2] with phi-squared coefficient and the second algorithm consists of minimizing the expression [4.2] with mutual information. But before developing these algorithms, we introduce a third distribution to facilitate the optimization of the criteria.

### 4.3.2. *Parameter modification of criteria*

Parameter modification [WIN 87, BRY 88] consists of increasing the number of parameters in a criterion so that the optimal value of the original parameters is not changed but is easier to optimize. The $k$-means algorithm is the most standard example using this principle: replacing the trace criterion $g(\mathbf{z}) = \sum_{i,k} d^2(\boldsymbol{x}_i, \mathbf{g}_k)$ with the criterion

$$\widetilde{g}(\mathbf{z}, \mathbf{a}) = \sum_{i,k} d^2(\boldsymbol{x}_i, \mathbf{a}_k)$$

with the new parameter $\mathbf{a} = (\mathbf{a}_1, \ldots, \mathbf{a}_g)$, the alternating optimization of this new criterion leads to the $k$-means algorithm.

Here, we introduce a new parameter $\boldsymbol{\delta} = (\delta_{k\ell})$, a matrix of size $(g, m)$ where each $\delta_{k\ell}$ plays the role of the centroid of the block $k\ell$ and such that

$$\delta_{k\ell} > 0 \quad \forall k, \ell \quad \text{and} \quad \sum_{k,\ell} p_{k.}^{\mathbf{z}} p_{.\ell}^{\mathbf{w}} \delta_{k\ell} = 1. \qquad [4.3]$$

We will note $\Delta(\mathbf{z}, \mathbf{w})$ the set of matrices $\delta$ which verify these constraints. Using this parameter, a new distribution $R_{IJ}^{\mathbf{zw}\delta} = (r_{ij}^{\mathbf{zw}\delta})$ depending on partitions $\mathbf{z}$ and $\mathbf{w}$ and parameter $\delta$ can be defined by

$$r_{ij}^{\mathbf{zw}\delta} = p_{i.}p_{.j} \sum_{k,\ell} z_{ik} w_{j\ell} \delta_{k\ell}.$$

The constraints [4.3] ensure that $R_{IJ}^{\mathbf{zw}\delta}$ is a distribution

$$\sum_{i,j} r_{ij}^{\mathbf{zw}\delta} = \sum_{i,j} p_{i.}p_{.j} \sum_{k,\ell} z_{ik} w_{j\ell} \delta_{k\ell} = \sum_{k,\ell} \delta_{k\ell} \sum_{i,j} p_{i.}p_{.j} z_{ik} w_{j\ell}$$
$$= \sum_{k,\ell} p_{k.}^{\mathbf{z}} p_{.\ell}^{\mathbf{w}} \delta_{k\ell} = 1.$$

This distribution $R_{IJ}^{\mathbf{zw}\delta}$ has also the same column and row margins that the distributions $P_{IJ}$ and $Q_{IJ}^{\mathbf{zw}}$

$$r_{i.}^{\mathbf{zw}\delta} = \sum_{j} r_{ij}^{\mathbf{zw}\delta} = \sum_{j} p_{i.}p_{.j} \sum_{k,\ell} z_{ik} w_{j\ell} \delta_{k\ell} = p_{i.} \sum_{j} p_{.j} = p_{i.}$$

and, symmetrically, $r_{.j}^{\mathbf{zw}\delta} = p_{.j}$.

Using these new distributions, the objective of co-clustering is replaced by minimizing the new criterion for the $\Phi^2$ measure of association

$$\widetilde{W}_{\Phi^2}(\mathbf{z}, \mathbf{w}, \boldsymbol{\delta}) = \Phi^2(P_{IJ}) - \Phi^2(R_{IJ}^{\mathbf{zw}\delta}) \qquad [4.4]$$

or the new criterion for the $\mathcal{I}$ measure of association

$$\widetilde{W}_{\mathcal{I}}(\mathbf{z}, \mathbf{w}, \boldsymbol{\delta}) = \mathcal{I}(P_{IJ}) - \mathcal{I}(R_{IJ}^{\mathbf{zw}\delta}). \qquad [4.5]$$

Finally, two equations which will be used in the next section can be established

$$\begin{aligned}\widetilde{W}_{\Phi^2}(\mathbf{z},\mathbf{w},\boldsymbol{\delta}) &= D_{\Phi^2}(P_{IJ}||R_{KL}^{\mathbf{zw}\boldsymbol{\delta}}) = \sum_{i,j}\frac{(p_{ij}-r_{ij})^2}{p_{i.}p_{.j}}\\
&= \sum_{i,j}\frac{(p_{ij}-p_{i.}p_{.j}\sum_{k,\ell}z_{ik}w_{j\ell}\delta_{k\ell})^2}{p_{i.}p_{.j}}\\
&= \sum_{i,j}p_{i.}p_{.j}\left(\frac{p_{ij}}{p_{i.}p_{.j}}-\sum_{k,\ell}z_{ik}w_{j\ell}\delta_{k\ell}\right)^2\\
&= \sum_{i,j}p_{i.}p_{.j}\left(\frac{(\sum_{k,\ell}z_{ik}w_{j\ell})p_{ij}}{p_{i.}p_{.j}}-\sum_{k,\ell}z_{ik}w_{j\ell}\delta_{k\ell}\right)^2\\
&= \sum_{i,j,k,\ell}z_{ik}w_{j\ell}p_{i.}p_{.j}\left(\frac{p_{ij}}{p_{i.}p_{.j}}-\delta_{k\ell}\right)^2,\end{aligned}\qquad [4.6]$$

and

$$\begin{aligned}\widetilde{W}_{\mathcal{I}}(\mathbf{z},\mathbf{w},\boldsymbol{\delta}) &= KL(P_{IJ}||R_{IJ}^{\mathbf{zw}\boldsymbol{\delta}}) = \sum_{i,j}p_{ij}\log\frac{p_{ij}}{r_{ij}}\\
&= \sum_{i,j}p_{ij}\log\frac{p_{ij}}{p_{i.}p_{.j}\sum_{k,\ell}z_{ik}w_{j\ell}\delta_{k\ell}}\\
&= \sum_{i,j}p_{ij}\log\frac{p_{ij}}{p_{i.}p_{.j}}-\sum_{i,j}p_{ij}\log\sum_{k,\ell}z_{ik}w_{j\ell}\delta_{k\ell} \qquad [4.7]\\
&= \sum_{i,j}p_{ij}\log\frac{p_{ij}}{p_{i.}p_{.j}}-\sum_{k\ell}\log\delta_{k\ell}\sum_{i,j}z_{ik}w_{ij}p_{ij}\\
&= \sum_{i,j}p_{ij}\log\frac{p_{ij}}{p_{i.}p_{.j}}-\sum_{k,\ell}p_{k\ell}^{\mathbf{zw}}\log\delta_{k\ell}.\end{aligned}$$

### 4.3.3. *Co-clustering with the phi-squared coefficient*

The maximization of the criterion $\widetilde{W}_{\Phi^2}(\mathbf{z}, \mathbf{w}, \boldsymbol{\delta})$ can be obtained by alternating the three computations: $\mathbf{z} = \arg\min_{\mathbf{z}} \widetilde{W}_{\Phi^2}(\mathbf{z}, \mathbf{w}, \boldsymbol{\delta})$, $\mathbf{w} = \arg\min_{\mathbf{w}} \widetilde{W}_{\Phi^2}(\mathbf{z}, \mathbf{w}, \boldsymbol{\delta})$ and $\boldsymbol{\delta} = \arg\min_{\boldsymbol{\delta}} \widetilde{W}_{\Phi^2}(\mathbf{z}, \mathbf{w}, \boldsymbol{\delta})$. Using equation [4.6], the criterion can be written

$$\widetilde{W}_{\Phi^2}(\mathbf{z}, \mathbf{w}, \boldsymbol{\delta}) = \sum_{i,k} z_{ik} p_{i.} \sum_{j,\ell} w_{j\ell} p_{.j} \left( \frac{p_{ij}}{p_{i.}p_{.j}} - \delta_{k\ell} \right)^2$$

and therefore, in the computation of $\mathbf{z}$, the minimization of $\widetilde{W}_{\Phi^2}(\mathbf{z}, \mathbf{w}, \boldsymbol{\delta})$ for fixed $\mathbf{w}$ and $\boldsymbol{\delta}$ is obtained by assigning each row $i$ to the cluster $k$ maximizing $\sum_{j,\ell} w_{j\ell} p_{.j} \left( \frac{p_{ij}}{p_{i.}p_{.j}} - \delta_{k\ell} \right)^2$. Similarly, in the computation of $\mathbf{w}$, the minimization of $\widetilde{W}_{\Phi^2}(\mathbf{z}, \mathbf{w}, \boldsymbol{\delta})$ for fixed $\mathbf{z}$ and $\boldsymbol{\delta}$ is obtained by assigning each column $j$ to the cluster $\ell$ maximizing $\sum_{i,k} z_{ik} p_{i.} \left( \frac{p_{ij}}{p_{i.}p_{.j}} - \delta_{k\ell} \right)^2$. Finally, for the computations of $\boldsymbol{\delta}$, using equation [4.6], the problem can be formulated as $\arg\min_{\delta_{k\ell}} F(\delta_{k\ell})$ for all $k, \ell$, where

$$\begin{aligned} F(\delta_{k\ell}) &= \sum_{i,j} z_{ik} w_{j\ell} p_{i.} p_{.j} \left( \frac{p_{ij}}{p_{i.}p_{.j}} - \delta_{k\ell} \right)^2 \\ &= p^{\mathbf{z}}_{k.} p^{\mathbf{w}}_{.\ell} \delta^2_{k\ell} - 2 p^{\mathbf{zw}}_{k\ell} \delta_{k\ell} + \sum_{i,j} z_{ik} w_{j\ell} \frac{p^2_{ij}}{p_{i.}p_{.j}} \end{aligned}$$

which yields to

$$\delta_{k\ell} = -\frac{-2 p^{\mathbf{zw}}_{k\ell}}{2(p^{\mathbf{z}}_{k.} p^{\mathbf{w}}_{.\ell})} = \frac{p^{\mathbf{zw}}_{k\ell}}{p^{\mathbf{z}}_{k.} p^{\mathbf{w}}_{.\ell}} \qquad \forall k, \ell.$$

Using the strategy described in section 2.4.1 of Chapter 2, we obtain the CROKI2 algorithm described in algorithm 4.1.

**Algorithm 4.1** CROKI2

**input:** x, $g$, $m$
**initialization:** z, w, $\delta_{k\ell} = \frac{p^{\mathbf{zw}}_{k\ell}}{p^{\mathbf{z}}_{k.}p^{\mathbf{w}}_{.\ell}}$
**repeat**
  **repeat**
    **step 1.** $z_i = \arg\min_k \sum_{j,\ell} w_{j\ell} p_{.j} (\frac{p_{ij}}{p_{i.}p_{.j}} - \delta_{k\ell})^2$
    **step 2.** $\delta_{k\ell} = \frac{p^{\mathbf{zw}}_{k\ell}}{p^{\mathbf{z}}_{k.}p^{\mathbf{w}}_{.\ell}}$
  **until** convergence
  **repeat**
    **step 3.** $w_j = \arg\min_\ell \sum_{i,k} z_{ik} p_{i.} (\frac{p_{ij}}{p_{i.}p_{.j}} - \delta_{k\ell})^2$
    **step 4.** $\delta_{k\ell} = \frac{p^{\mathbf{zw}}_{k\ell}}{p^{\mathbf{z}}_{k.}p^{\mathbf{w}}_{.\ell}}$
  **until** convergence
**until** convergence
**return** z, w, $\delta$

This algorithm defines a sequence $(\mathbf{z}^{(t)}, \mathbf{w}^{(t)}, \boldsymbol{\delta}^{(t)})$ which monotonically decreases the criterion $\widetilde{\mathrm{W}}_{\Phi^2}(\mathbf{z}^{(t)}, \mathbf{w}^{(t)}, \boldsymbol{\delta}^{(t)})$ and, as the parameter is equal to $\delta^{(t)}_{k\ell} = \frac{(p^{\mathbf{zw}}_{k\ell})^{(t)}}{(p^{\mathbf{z}}_{k.})^{(t)}(p^{\mathbf{w}}_{.\ell})^{(t)}}$, we obtain

$$\begin{aligned}\widetilde{\mathrm{W}}_{\Phi^2}(\mathbf{z}^{(t)}, \mathbf{w}^{(t)}, \boldsymbol{\delta}^{(t)}) &= \sum_{i,j,k,\ell} z^{(t)}_{ik} w^{(t)}_{j\ell}\, p_{i.}p_{.j} \left(\frac{p_{ij}}{p_{i.}p_{.j}} - \delta^{(t)}_{k\ell}\right)^2 \\ &= \sum_{i,j,k,\ell} z^{(t)}_{ik} w^{(t)}_{j\ell}\, p_{i.}p_{.j} \left(\frac{p_{ij}}{p_{i.}p_{.j}} - \frac{(p^{\mathbf{zw}}_{k\ell})^{(t)}}{(p^{\mathbf{z}}_{k.})^{(t)}(p^{\mathbf{w}}_{.\ell})^{(t)}}\right)^2 \\ &= \sum_{i,j} \frac{(p_{ij} - (q^{\mathbf{zw}}_{ij})^{(t)})^2}{p_{i.}p_{.j}} = W_{\Phi^2}(\mathbf{z}^{(t)}, \mathbf{w}^{(t)})\end{aligned}$$

which proves that this algorithm also monotonically decreases the initial criterion $W_{\Phi^2}(\mathbf{z}^{(t)}, \mathbf{w}^{(t)})$. It can be shown that it converges to a local optimum of the criterion. CROKI2 is computationally efficient even for sparse data and its complexity can be shown to be $O(nz * it * (g + m))$ where $nz$ is the number of non-zero values in the input contingency table and $it$ the number of iterations. Empirically, 20 iterations are seen to suffice.

Hereafter, we propose to illustrate CROKI2 by applying it to time-budget data sets used in Chapter 1. Let us recall that we have a data matrix of size $28 \times 10$ where $x_{ij}$ represents the amount of time spent on a variety of activities $i$ by a population $j$ during a given time period. The initial $\Phi^2(I, J)$ value is 0.14392 and the resulting $\Phi^2(\mathrm{z}, \mathrm{w})$ value is 0.11993. The percentage of $\Phi^2$ accounted for by the co-clustering is very good in this example: more than 83% of $\Phi^2$ is preserved. The clusters are the following:

– Row partition:

- waus wcus wawe wcwe wcyo wces;

- wayo waes wmes;

- wmus wmwe wmyo;

- wnau wnaw wnay wnae;

- maus mmus mcus mawe mmwe mcwe may0 mmyo mcyo maes mmes mces.

– Column partition:

- home child;

- prof tran;

- shop wash meal sleap tv leis.

and the reorganized data matrix is presented in Table 4.4. In Figure 4.5, we present data matrix $\mathbf{x}^{\mathrm{zw}}$ and its associated data matrix $\mathbf{P}^{\mathrm{zw}} = (\frac{p_{k\ell}}{p_{k.}p_{.\ell}}) \times 1,000$. With a multiple coefficient $p_{k.}p_{.\ell}$, each cell $\mathbf{P}^{\mathrm{zw}}$ measures the deviance between the centroid of a block and the centroid of the initial data that is equal to 1. The most interesting values are therefore those which are far from the mean 1 (that is 1000 in our case).

These values characterize the 15 blocks. For instance, column clusters 1 and 2 are features of row clusters 1, 4 and 5. Row cluster 1 has a high value for column cluster 1 and a small value for column cluster 2, while row clusters 4 and 5 have high values for column cluster 2 and small values for cluster 1. Note that reciprocally these row clusters are characteristic of these column clusters. At the same time, we note that row cluster 3 and column cluster 3 do not have any characteristics; the values are close to 1,000. This observation is confirmed by looking at the planes obtained by CA in Figures 4.2 and 4.3. The clusters are located near the center of gravity. Furthermore, the simultaneous visualization of planes confirms the opposition of row cluster 1 and row clusters 4 and 5 according to column clusters 1 and 2.

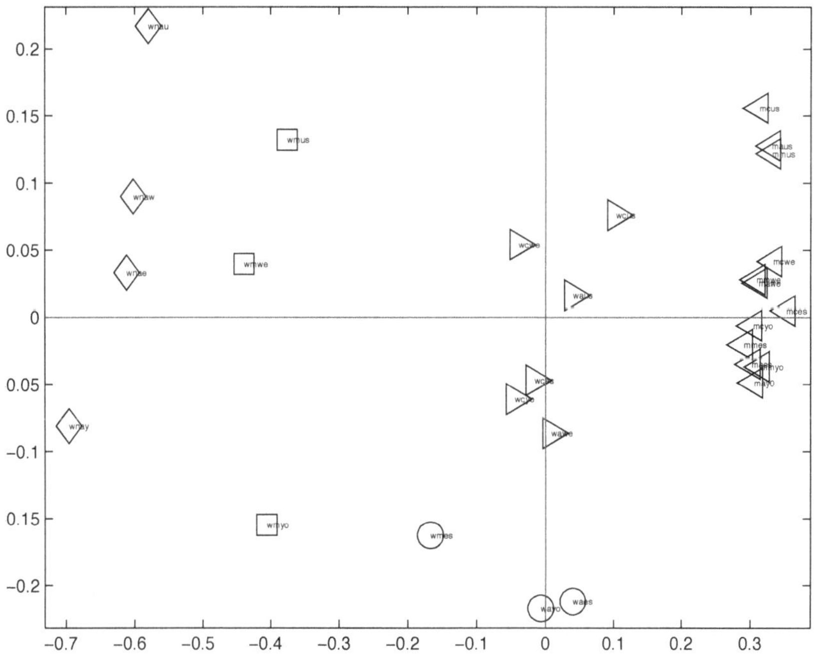


**Figure 4.2.** *Projection of the columns into the factorial plane spawned by the first and second axes that account for 84% of* $\phi^2$

| | home | child | prof | tran | shop | wash | meal | sleap | tv | leis |
|---|---|---|---|---|---|---|---|---|---|---|
| waus | 250 | 30 | 475 | 90 | 140 | 120 | 100 | 775 | 115 | 305 |
| wcus | 196 | 18 | 482 | 94 | 141 | 130 | 96 | 775 | 132 | 336 |
| wawe | 307 | 30 | 510 | 70 | 80 | 95 | 142 | 815 | 87 | 262 |
| wcwe | 262 | 14 | 389 | 34 | 92 | 97 | 147 | 848 | 84 | 392 |
| wcyo | 318 | 23 | 413 | 89 | 112 | 96 | 102 | 774 | 45 | 409 |
| wces | 296 | 21 | 433 | 86 | 128 | 102 | 94 | 758 | 58 | 379 |
| wayo | 375 | 45 | 560 | 105 | 90 | 90 | 95 | 745 | 60 | 235 |
| waes | 338 | 42 | 578 | 106 | 106 | 94 | 52 | 752 | 64 | 228 |
| wmes | 431 | 60 | 434 | 77 | 117 | 88 | 105 | 770 | 73 | 229 |
| wmus | 421 | 87 | 179 | 29 | 161 | 112 | 119 | 776 | 143 | 373 |
| wmwe | 529 | 69 | 168 | 22 | 102 | 83 | 174 | 825 | 119 | 392 |
| wmyo | 576 | 59 | 260 | 52 | 116 | 85 | 117 | 775 | 65 | 295 |
| wnau | 495 | 110 | 10 | 0 | 170 | 110 | 130 | 785 | 160 | 430 |
| wnaw | 567 | 87 | 20 | 7 | 112 | 90 | 180 | 842 | 125 | 367 |
| wnay | 710 | 55 | 10 | 10 | 145 | 85 | 130 | 815 | 60 | 380 |
| wnae | 594 | 72 | 24 | 8 | 158 | 92 | 128 | 840 | 86 | 398 |
| maus | 60 | 10 | 610 | 140 | 120 | 95 | 115 | 760 | 175 | 315 |
| mmus | 65 | 10 | 615 | 141 | 115 | 90 | 115 | 765 | 180 | 305 |
| mcus | 50 | 0 | 585 | 115 | 150 | 105 | 100 | 760 | 150 | 385 |
| mawe | 95 | 7 | 652 | 100 | 57 | 85 | 150 | 807 | 115 | 330 |
| mmwe | 97 | 10 | 655 | 97 | 52 | 85 | 152 | 807 | 122 | 320 |
| mcwe | 72 | 0 | 642 | 105 | 62 | 77 | 140 | 812 | 100 | 387 |
| may0 | 120 | 15 | 650 | 140 | 85 | 90 | 105 | 760 | 70 | 365 |
| mmyo | 112 | 15 | 650 | 145 | 85 | 90 | 105 | 760 | 80 | 357 |
| mcyo | 95 | 0 | 615 | 125 | 115 | 90 | 85 | 760 | 40 | 475 |
| maes | 122 | 22 | 650 | 142 | 76 | 94 | 100 | 764 | 96 | 334 |
| mmes | 134 | 22 | 652 | 133 | 68 | 94 | 102 | 762 | 122 | 310 |
| mces | 68 | 0 | 627 | 148 | 88 | 92 | 86 | 770 | 58 | 463 |

**Table 4.4.** *Co-clustering obtained with* CROKI2 *on a time-budget data set*

Finally, in Figure 4.4, a synthetic representation of initial and reorganized data, and the reorganized averaged data are shown.

| | 1 | 2 | 3 |
|---|---|---|---|
| 1 | 1765 | 3165 | 9363 |
| 2 | 1291 | 1860 | 3993 |
| 3 | 1741 | 710 | 4832 |
| 4 | 2690 | 89 | 6818 |
| 5 | 1201 | 9134 | 18456 |

| | 1 | 2 | 3 |
|---|---|---|---|
| 1 | 954 | 993 | 1011 |
| 2 | 1396 | 1168 | 863 |
| 3 | 1846 | 437 | 1024 |
| 4 | 2165 | 42 | 1097 |
| 5 | 322 | 1423 | 990 |

**Table 4.5.** $\mathbf{x}^{\mathbf{zw}}$: *Summary of initial data matrix and* $\mathbf{P}^{\mathbf{zw}}$: *deviance between each centroid and 1*

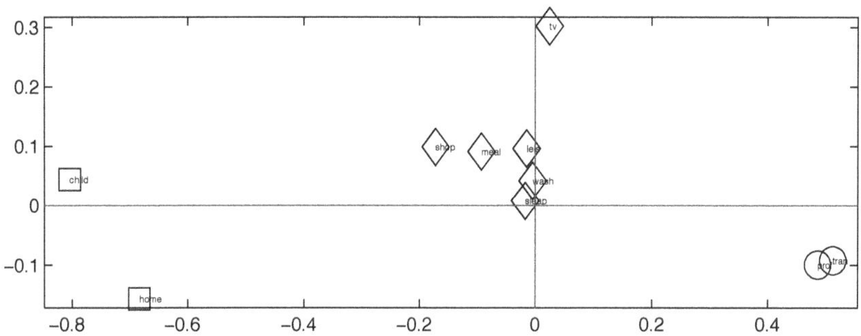


**Figure 4.3.** *Projection of the columns into the factorial plane spawned by the first and second axes that account for 84% of* $\phi^2$

### 4.3.4. *Co-clustering with the mutual information*

The maximization of the criterion $\widetilde{W}_{\mathcal{I}}(\mathbf{z}, \mathrm{w}, \boldsymbol{\delta})$ can be obtained by alternating the three computations: $\mathbf{z} = \arg\min_{\mathbf{z}} \widetilde{W}_{\mathcal{I}}(\mathbf{z}, \mathrm{w}, \boldsymbol{\delta})$, $\mathrm{w} = \arg\min_{\mathrm{w}} \widetilde{W}_{\mathcal{I}}(\mathbf{z}, \mathrm{w}, \boldsymbol{\delta})$ and $\boldsymbol{\delta} = \arg\min_{\boldsymbol{\delta}} \widetilde{W}_{\mathcal{I}}(\mathbf{z}, \mathrm{w}, \boldsymbol{\delta})$. Using equation [4.7], the criterion can be written as

$$\widetilde{W}_{\mathcal{I}}(\mathbf{z}, \mathrm{w}, \boldsymbol{\delta}) = \sum_{i,j} p_{ij} \log \frac{p_{ij}}{p_{i.}p_{.j}} - \sum_{i,k} z_{ik} \sum_{\ell} p_{i\ell}^{\mathrm{w}} \log \delta_{k\ell}$$

and therefore, in the computation of z, the minimization of $\widetilde{W}_{\mathcal{I}}(\mathbf{z}, \mathrm{w}, \boldsymbol{\delta})$ for fixed w and $\boldsymbol{\delta}$ is obtained by assigning each row $i$ to the cluster $k$ maximizing

$$\sum_{\ell} (\sum_{j} w_{j\ell} p_{ij}) \log \delta_{k\ell}.$$

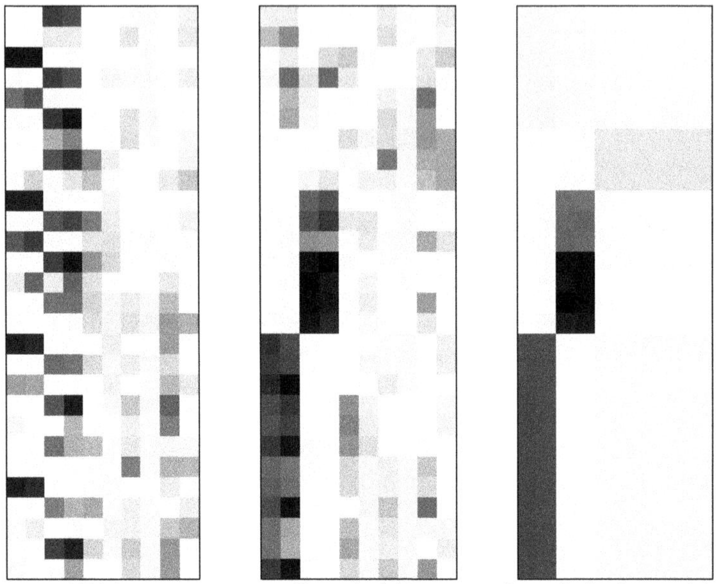


**Figure 4.4.** *Visualization of initial data and results after co-clustering*

Similarly, in the computation of w, the minimization of $\widetilde{\mathrm{W}}_{\mathcal{I}}(\mathrm{z}, \mathrm{w}, \delta)$ for fixed z and $\delta$ is obtained by assigning each column $j$ to the cluster $\ell$ maximizing

$$\sum_k (\sum_i z_{ik} p_{ij}) \log \delta_{k\ell}.$$

Finally, for the computations of $\delta$, the problem can be formulated as

$$\arg\max_{\delta} \sum_{k,\ell} p^{\mathbf{zw}}_{k\ell} \log \delta_{k\ell}$$

with $\sum_{k,\ell} p^{\mathbf{z}}_{k.} p^{\mathbf{w}}_{.\ell} \delta_{k\ell} = 1$ which yields to $\delta_{k\ell} = \frac{p^{\mathbf{zw}}_{k\ell}}{p^{\mathbf{z}}_{k.} p^{\mathbf{w}}_{.\ell}}$ for all $k, \ell$. A strategy of alternated minimization similar to the CROKI2 algorithm leads to the CROINFO algorithm (algorithm 4.2) .

**Algorithm 4.2** CROINFO

**input:** x, $g$, $m$
**initialization:** z, w, $\delta_{k\ell} = \frac{p^{\mathbf{zw}}_{k\ell}}{p^{\mathbf{z}}_{k.}p^{\mathbf{w}}_{.\ell}}$
**repeat**
  **repeat**
    **step 1.** $z_i = \arg\min_k \sum_\ell p^{\mathbf{w}}_{i\ell} \log \delta_{k\ell}$
    **step 2.** $\delta_{k\ell} = \frac{p^{\mathbf{zw}}_{k\ell}}{p^{\mathbf{z}}_{k.}p^{\mathbf{w}}_{.\ell}}$
  **until** convergence
  **repeat**
    **step 3.** $w_j = \arg\min_\ell \sum_k p^{\mathbf{z}}_{kj} \log \delta_{k\ell}$
    **step 4.** $\delta_{k\ell} = \frac{p^{\mathbf{zw}}_{k\ell}}{p^{\mathbf{z}}_{k.}p^{\mathbf{w}}_{.\ell}}$
  **until** convergence
**until** convergence
**return** z and w

It can also be shown that this algorithm monotonically decreases the criterion $W_{\mathcal{I}}(\mathbf{z}^{(t)}, \mathbf{w}^{(t)})$ and its convergence properties are the same as the properties of the CROKI2 algorithm.

To illustrate the CROINFO algorithm, we present in Table 4.6 the distribution $R^{\mathbf{zw}\delta}_{IJ}$ obtained after the different steps of this algorithm are applied to the contingency table of Table 4.1. In Table 4.7, the values of phi-squared coefficient and mutual information of the initial distribution $P_{IJ}$ and final distribution $Q^{\mathbf{zw}}_{IJ}$ are presented, and the corresponding criterion W expressing a loss information. It can be observed that this algorithm gives progressively better clustering and approximations to reach a loss information equal to 0.04.

## 4.4. Model-based co-clustering

Model-based clustering assumes that the data were generated by a model and tries to recover the original model from the data. For the co-clustering problem, these models (see Chapter 2) assumed that there exists a partition z into $g$

clusters on $I$ and a partition w into $m$ clusters on $J$, such that the random variables $x_{ij}$ are conditionally independent knowing z and w. Two approaches can be considered:

After step z (iteration 1)

| $R^{zw\delta}_{IJ}$ | 1 | 2 | 3 | 4 | 5 | z |
|---|---|---|---|---|---|---|
| 1 | 0.035 | 0.030 | 0.029 | 0.034 | 0.032 | 2 |
| 2 | 0.037 | 0.032 | 0.027 | 0.035 | 0.030 | 1 |
| 3 | 0.029 | 0.025 | 0.027 | 0.028 | 0.030 | 3 |
| 4 | 0.029 | 0.025 | 0.023 | 0.027 | 0.026 | 2 |
| 5 | 0.046 | 0.040 | 0.038 | 0.044 | 0.042 | 2 |
| 6 | 0.042 | 0.036 | 0.039 | 0.040 | 0.043 | 3 |
| w | 2 | 2 | 1 | 2 | 1 | 1 |

After step z (iteration 2)

| $R^{zw\delta}_{IJ}$ | 1 | 2 | 3 | 4 | 5 | z |
|---|---|---|---|---|---|---|
| 1 | 0.056 | 0.048 | 0.046 | 0.005 | 0.005 | 1 |
| 2 | 0.056 | 0.048 | 0.046 | 0.005 | 0.005 | 1 |
| 3 | 0.023 | 0.020 | 0.019 | 0.040 | 0.038 | 3 |
| 4 | 0.021 | 0.018 | 0.017 | 0.037 | 0.035 | 3 |
| 5 | 0.045 | 0.039 | 0.037 | 0.045 | 0.043 | 2 |
| 6 | 0.043 | 0.037 | 0.035 | 0.043 | 0.041 | 2 |
| w | 2 | 2 | 2 | 1 | 1 | 1 |

After step $\delta$ (iteration 1)

| $R^{zw\delta}_{IJ}$ | 1 | 2 | 3 | 4 | 5 | z |
|---|---|---|---|---|---|---|
| 1 | 0.035 | 0.030 | 0.029 | 0.034 | 0.032 | 2 |
| 2 | 0.039 | 0.034 | 0.024 | 0.037 | 0.026 | 1 |
| 3 | 0.029 | 0.025 | 0.027 | 0.028 | 0.030 | 3 |
| 4 | 0.029 | 0.025 | 0.023 | 0.027 | 0.026 | 2 |
| 5 | 0.046 | 0.040 | 0.038 | 0.044 | 0.042 | 2 |
| 6 | 0.042 | 0.036 | 0.039 | 0.040 | 0.043 | 3 |
| w | 2 | 2 | 1 | 2 | 1 | 1 |

After step $\delta$ (iteration 2)

| $R^{zw\delta}_{IJ}$ | 1 | 2 | 3 | 4 | 5 | z |
|---|---|---|---|---|---|---|
| 1 | 0.056 | 0.048 | 0.046 | 0.005 | 0.005 | 1 |
| 2 | 0.056 | 0.048 | 0.046 | 0.005 | 0.005 | 1 |
| 3 | 0.008 | 0.007 | 0.006 | 0.061 | 0.058 | 3 |
| 4 | 0.007 | 0.006 | 0.006 | 0.057 | 0.054 | 3 |
| 5 | 0.048 | 0.041 | 0.039 | 0.042 | 0.040 | 2 |
| 6 | 0.045 | 0.039 | 0.037 | 0.040 | 0.038 | 2 |
| w | 2 | 2 | 2 | 1 | 1 | 1 |

After step w (iteration 1)

| $R^{zw\delta}_{IJ}$ | 1 | 2 | 3 | 4 | 5 | z |
|---|---|---|---|---|---|---|
| 1 | 0.035 | 0.030 | 0.029 | 0.034 | 0.032 | 2 |
| 2 | 0.039 | 0.034 | 0.032 | 0.028 | 0.026 | 1 |
| 3 | 0.029 | 0.025 | 0.024 | 0.032 | 0.030 | 3 |
| 4 | 0.029 | 0.025 | 0.023 | 0.027 | 0.026 | 2 |
| 5 | 0.046 | 0.040 | 0.038 | 0.044 | 0.042 | 2 |
| 6 | 0.042 | 0.036 | 0.034 | 0.046 | 0.043 | 3 |
| w | 2 | 2 | 2 | 1 | 1 | 1 |

After step w (iteration 2)

| $R^{zw\delta}_{IJ}$ | 1 | 2 | 3 | 4 | 5 | z |
|---|---|---|---|---|---|---|
| 1 | 0.056 | 0.048 | 0.046 | 0.005 | 0.005 | 1 |
| 2 | 0.056 | 0.048 | 0.046 | 0.005 | 0.005 | 1 |
| 3 | 0.008 | 0.007 | 0.006 | 0.061 | 0.058 | 3 |
| 4 | 0.007 | 0.006 | 0.006 | 0.057 | 0.054 | 3 |
| 5 | 0.048 | 0.041 | 0.039 | 0.042 | 0.040 | 2 |
| 6 | 0.045 | 0.039 | 0.037 | 0.040 | 0.038 | 2 |
| w | 2 | 2 | 2 | 1 | 1 | 1 |

After step $\delta$ (iteration 1)

| $R^{zw\delta}_{IJ}$ | 1 | 2 | 3 | 4 | 5 | z |
|---|---|---|---|---|---|---|
| 1 | 0.035 | 0.030 | 0.028 | 0.034 | 0.033 | 2 |
| 2 | 0.056 | 0.048 | 0.046 | 0.005 | 0.005 | 1 |
| 3 | 0.023 | 0.020 | 0.019 | 0.040 | 0.038 | 3 |
| 4 | 0.028 | 0.024 | 0.023 | 0.028 | 0.027 | 2 |
| 5 | 0.045 | 0.039 | 0.037 | 0.045 | 0.043 | 2 |
| 6 | 0.033 | 0.028 | 0.027 | 0.057 | 0.055 | 3 |
| w | 2 | 2 | 2 | 1 | 1 | 1 |

After step $\delta$ (iteration 2)

| $R^{zw\delta}_{IJ}$ | 1 | 2 | 3 | 4 | 5 | z |
|---|---|---|---|---|---|---|
| 1 | 0.056 | 0.048 | 0.046 | 0.005 | 0.005 | 1 |
| 2 | 0.056 | 0.048 | 0.046 | 0.005 | 0.005 | 1 |
| 3 | 0.008 | 0.007 | 0.006 | 0.061 | 0.058 | 3 |
| 4 | 0.007 | 0.006 | 0.006 | 0.057 | 0.054 | 3 |
| 5 | 0.048 | 0.041 | 0.039 | 0.042 | 0.040 | 2 |
| 6 | 0.045 | 0.039 | 0.037 | 0.040 | 0.038 | 2 |
| w | 2 | 2 | 2 | 1 | 1 | 1 |

**Table 4.6.** *First two iterations of* CROINFO

| | $\Phi^2$ | $\mathcal{I}$ |
|---|---|---|
| $P_{IJ}$ | 0.415 | 0.254 |
| $P^{zw}_{KL}$ or $Q^{zw}_{IJ}$ | 0.378 | 0.214 |
| W | 0.037 | 0.040 |

**Table 4.7.** *Phi-squared coefficient and mutual information of initial distribution $P_{IJ}$ and final distribution $Q^{zw}_{IJ}$, and corresponding loss criterion $W$*

– In the first approach, the most common, the data are considered as a sample of size $N$ of a bivariate categorical random variable with $n$ and $d$ levels and the contingency table x is an $n \times d$ matrix in which the cells contain frequency counts of the $n \times d$ possible outcomes. Therefore, the contingency table has the multinomial distribution $\mathcal{M}(N, p_{11}, \ldots, p_{nd})$ characterized by the probability $p_{ij}$ of each cell. In this situation, all probabilistic models consist of giving the form of the probabilities $p_{ij}$, and the best-known ones for contingency tables are the log-linear models. The model proposed here, the *block model*, used the partitions of the levels of the two categorical variables as parameters of the model.

– In the second approach, co-occurrence data are defined in two samples (authors and publications, words and documents, and so on) of size $n$ and $d$ which, unlike the first approach are not fixed. Therefore, it is difficult to consider that the partitions, the dimensions of which will vary with the size of the data, are model parameters. The model proposed here, LBM, solves this problem by considering that the partitions are latent variables.

In this section, we examine these two approaches and show the links with the algorithms described in the previous section.

### 4.4.1. *Block model for contingency tables*

#### 4.4.1.1. *Definition of the model*

This model considers that the contingency table x is distributed according to a multinomial distribution with parameters $(N, \xi_{11}, \ldots, \xi_{nd})$ where cell probabilities $\xi_{ij}$ of the $I \times J$ contingency table take the following form

$$\xi_{ij} = \alpha_i \beta_j \sum_{k,\ell} z_{ik} w_{j\ell} \delta_{k\ell}$$

for a row effect $\alpha_i$, a column effect $\beta_j$ and a block effect $\delta_{k\ell}$. Its pdf takes the following form

$$f(\mathbf{x};\boldsymbol{\theta}) = \frac{N!}{\prod_{i,j} x_{ij}!}\prod_{i,j}\xi_{ij}^{x_{ij}}$$

$$= \frac{N!}{\prod_{i,j} x_{ij}!}\prod_{i,j}\left(\alpha_i\beta_j\sum_{k,\ell} z_{ik}w_{j\ell}\delta_{k\ell}\right)^{x_{ij}}, \qquad [4.8]$$

where $\boldsymbol{\theta} = (\boldsymbol{\alpha},\boldsymbol{\beta},\boldsymbol{\delta},\mathbf{z},\mathbf{w})$ with $\boldsymbol{\alpha} = (\alpha_1,\ldots,\alpha_n)$, $\boldsymbol{\beta} = (\beta_1,\ldots,\beta_d)$, $\boldsymbol{\delta} = (\delta_{11},\ldots,\delta_{gm})$ and $\mathbf{z}$ and $\mathbf{w}$ are partitions of the rows $I$ and columns $J$. As for the log-linear models for contingency tables, identifiability requires constraints. Denoting $\alpha_k = \sum_i z_{ik}\alpha_i$ and $\beta_\ell = \sum_j w_{j\ell}\beta_j$, the constraints

$$\sum_i \alpha_i = \sum_j \beta_j = \sum_k \alpha_k\delta_{k\ell} = \sum_\ell \beta_\ell\delta_{k\ell} = 1 \quad \forall k,\ell,$$

which ensure that $\sum_{i,j}\xi_{ij} = 1$, will be chosen.

4.4.1.2. *Maximum likelihood estimation*

The maximum likelihood estimates are parameter values that maximize the likelihood [4.8] or the log-likelihood which can be written as

$$L(\mathbf{z},\mathbf{w},\boldsymbol{\alpha},\boldsymbol{\beta},\boldsymbol{\delta}) = \sum_{i,j} x_{ij}\log\left(\alpha_i\beta_j\sum_{k,\ell} z_{ik}w_{j\ell}\delta_{k\ell}\right) + C$$

$$= \sum_i x_{i.}\log\alpha_i + \sum_j x_{.j}\log\beta_j$$

$$+\sum_{k,\ell} x_{k\ell}^{\mathbf{zw}}\log\delta_{k\ell} + C$$

where $C$ does not depend on z, w, $\alpha$, $\beta$ and $\delta$. It can be easily shown that $\alpha_i = p_{i.}$, $\beta_j = p_{.j}$ and the estimation of the parameters z, w and $\delta$ are obtained by maximization of

$$\sum_{k,\ell} x_{k\ell}^{\mathbf{zw}} \log \delta_{k\ell} = N \sum_{k,\ell} p_{k\ell}^{\mathbf{zw}} \log \delta_{k\ell}$$

which is exactly the problem of co-clustering with the mutual information discussed in section 4.3.4, formulated as

$$\arg\min_{\mathbf{z},\mathbf{w},\boldsymbol{\delta}} \widetilde{W}_{\mathcal{I}}(\mathbf{z}, \mathbf{w}, \boldsymbol{\delta}) = \arg\max_{\mathbf{z},\mathbf{w},\boldsymbol{\delta}} \sum_{k,\ell} p_{k\ell}^{\mathbf{zw}} \log \delta_{k\ell}$$

and solved by the CROINFO algorithm. Thus, we give a probabilistic interpretation of the minimization problem of mutual information loss function by this algorithm. Note that we have used multinomial distribution and assumed $N$ to be known. In certain situations, $N$ can be unknown and in that case, we can use a Poisson distribution for counts of events that occur randomly. With this distribution, the same calculus can easily be performed and leads to the same conclusion.

4.4.1.3. *Minimum chi-squared estimation*

In the minimum chi-squared estimation [HAR 83], which is an alternative to maximum likelihood estimation, the parameter estimations are obtained by minimizing

$$A_1 = N \sum_{i,j} \frac{(p_{ij} - \xi_{ij})^2}{\xi_{ij}}.$$

The expected frequency $N\xi_{ij}$ in the denominator causes certain difficulties and a modification has been suggested which makes the computation easier. For instance, the modified chi-squared statistic proposed by Neyman [NEY 49]

$$A_2 = N \sum_{i,j} \frac{(p_{ij} - \xi_{ij})^2}{p_{ij}}$$

leads to estimators that have the same properties as those provided by the minimum chi-squared. Another solution is to use the modified chi-squared statistic

$$A_3 = N \sum_{i,j} \frac{(p_{ij} - \xi_{ij})^2}{p_{i.}p_{.j}}.$$

No theoretical property seems to have been proved but a very good approximation of the expression $A_1$ obtained by $A_3$ (much better than that obtained by $A_2$) on many examples led us to use this expression to establish the minimum chi-squared estimation.

Unfortunately, the estimation of row and column effects is difficult. One solution is to estimate these effects by their sufficient statistics. Under the multinomial model, the sufficient statistic is expressible as a linear combination of the corresponding cell proportions, and therefore a sufficient statistic of $\xi_{i.}$ is $p_{i.} = \frac{x_{i.}}{N}$ and, as we have

$$\xi_{i.} = \sum_{j} \alpha_i \beta_j \sum_{k,\ell} z_{ik} w_{j\ell} \delta_{k\ell} = \alpha_i \sum_{k} z_{ik} \underbrace{\sum_{\ell} \beta_\ell \delta_{k\ell}}_{1} = \alpha_i,$$

the $p_{i.}$s are sufficient statistics of row effects $\alpha_i$. In a similar way, we obtain that the $p_{.j}$s are sufficient statistics of row effects $\beta_j$. Therefore, the estimations of parameters z, w and $\delta$ are obtained by minimizing

$$\sum_{i,j} \frac{(x_{ij} - Np_{i.}p_{.j} \sum_{k\ell} z_{ik} w_{j\ell} \delta_{k\ell})^2}{Np_{i.}p_{.j}} = N \sum_{i,j} \frac{(p_{ij} - r_{ij}^{\mathbf{zw}\delta})^2}{p_{i.}p_{.j}}$$

which is equal, up to the multiplicative constant $N$, to the criterion $\widetilde{W}_{\Phi^2}(\mathbf{z}, \mathrm{w}, \delta)$ and therefore the estimation can be solved by the CROKI2 algorithm.

Finally, the two approaches of co-clustering of a contingency table defined in the previous section clustering can be viewed as estimating the parameters of a probabilistic block model by two different estimation methods.

### 4.4.2. *Poisson latent block model*

#### 4.4.2.1. *The model*

Using the LBM described in section 2.3 of Chapter 2 in the contingency table situation and assuming that for each block $k\ell$ the values $x_{ij}$ are distributed according to the Poisson distribution $\mathcal{P}(\lambda_{ij})$ where the parameters $\lambda_{ij}$ take the following form

$$\lambda_{ij} = \mu_i \nu_j \sum_{k,\ell} z_{ik} w_{j\ell} \gamma_{k\ell}$$

for a row effect $\mu_i$, a column effect $\nu_j$ and a block effect $\gamma_{k\ell}$, we obtain the Poisson LBM [GOV 10] with the following pdf

$$f(\mathbf{x};\boldsymbol{\theta}) = \sum_{(\mathbf{z},\mathbf{w})\in\mathcal{Z}\times\mathcal{W}} \prod_{i,k} \pi_k^{z_{ik}} \prod_{j,\ell} \rho_\ell^{w_{j\ell}} \times \prod_{i,j} \frac{e^{-\mu_i\nu_j \sum_{k,\ell} z_{ik}w_{j\ell}\gamma_{k\ell}} (\mu_i\nu_j \sum_{k,\ell} z_{ik}w_{j\ell}\gamma_{k\ell})^{x_{ij}}}{x_{ij}!} \qquad [4.9]$$

parameterized by $\boldsymbol{\theta} = (\boldsymbol{\pi}, \boldsymbol{\rho}, \boldsymbol{\mu}, \boldsymbol{\nu}, \boldsymbol{\gamma})$ with $\boldsymbol{\pi} = (\pi_1, \ldots, \pi_g)$, $\boldsymbol{\rho} = (\rho_1, \ldots, \rho_m)$, $\boldsymbol{\mu} = (\mu_1, \ldots, \mu_n)$, $\boldsymbol{\nu} = (\nu_1, \ldots, \nu_d)$ and $\boldsymbol{\gamma} = (\gamma_{11}, \ldots, \gamma_{gm})$.

As for the previous block models, identifiability requires constraints. Denoting $\mu_k = \sum_i z_{ik}\mu_i$ and $\nu_\ell = \sum_j w_{j\ell}\nu_j$, the constraints

$$\sum_i \mu_i = \sum_j \nu_j = 1/\sum_k \pi_k \gamma_{k\ell} = 1/\sum_\ell \rho_\ell \gamma_{k\ell} = M \quad \forall k, \ell$$

are chosen. Under these constraints, it can be shown that $\mathbb{E}(x_{i.}) = \mu_i$ and $\mathbb{E}(x_{.j}) = \nu_j$ and $\mu_i$ and $\nu_j$ can be replaced by the margins $x_{i.}$ and $x_{.j}$; the parameters to be estimated are therefore reduced to $\boldsymbol{\pi}$, $\boldsymbol{\rho}$ and $\boldsymbol{\gamma}$. With this model, the complete-data log-likelihood is given, up to a constant, by

$$\begin{aligned}\mathrm{L_C}(\mathbf{z},\mathbf{w},\boldsymbol{\theta}) &= \sum_k z_{.k}\log\pi_k + \sum_\ell w_{.\ell}\log\rho_\ell \\ &\quad + \sum_{i,j,k,\ell} z_{ik}w_{j\ell}(x_{ij}\log\gamma_{k\ell} - x_{i.}x_{.j}\gamma_{k\ell}).\end{aligned}$$

### 4.4.3. *Poisson* LBVEM *and* LBCEM *algorithms*

Taking into account the definition of the pdf $f(x_{ij};\alpha_{k\ell})$, the variational approximation of the EM algorithm described in section 2.4.1 of Chapter 2 is obtained by alternating the three computations $\arg\max_{\widetilde{\mathbf{z}}} F_C(\widetilde{\mathbf{z}},\widetilde{\mathbf{w}},\boldsymbol{\theta})$, $\arg\max_{\widetilde{\mathbf{w}}} F_C(\widetilde{\mathbf{z}},\widetilde{\mathbf{w}},\boldsymbol{\theta})$ and $\arg\max_{\boldsymbol{\theta}} F_C(\widetilde{\mathbf{z}},\widetilde{\mathbf{w}},\boldsymbol{\theta})$.

– Computation of $\boldsymbol{\theta}$: we obtain $\pi_k = \frac{\widetilde{z}_{.k}}{n}$, $\rho_\ell = \frac{\widetilde{w}_{.\ell}}{n}$ and $\gamma_{k\ell}$ is defined as follows

$$\begin{aligned}\gamma_{k\ell} &= \arg\max_{\gamma_{k\ell}} \sum_{i,j} \widetilde{z}_{ik}\widetilde{w}_{j\ell}(x_{ij}\log\gamma_{k\ell} - x_{i.}x_{.j}\gamma_{k\ell}) \\ &= \arg\max_{\gamma_{k\ell}} \left(x^{\widetilde{\mathbf{z}}\widetilde{\mathbf{w}}}_{k\ell} - x^{\widetilde{\mathbf{z}}}_{k.}x^{\widetilde{\mathbf{w}}}_{.\ell}\right)\end{aligned}$$

and therefore $\gamma_{k\ell} = \frac{x^{\widetilde{\mathbf{z}}\widetilde{\mathbf{w}}}_{k\ell}}{x^{\widetilde{\mathbf{z}}}_{k.}x^{\widetilde{\mathbf{w}}}_{.\ell}}$.

– Computation of $\widetilde{\mathbf{z}}$: we obtain

$$\begin{aligned}\widetilde{z}_{ik} = \arg\max_{\widetilde{z}_{ik}} \sum_k \Bigg(&\widetilde{z}_{ik}\log\pi_k \\ &+\widetilde{z}_{ik}\sum_\ell (x^{\widetilde{\mathbf{w}}}_{i\ell}\log\gamma_{k\ell} - x_{i.}x^{\widetilde{\mathbf{w}}}_{.\ell}\gamma_{k\ell}) - \widetilde{z}_{ik}\log\widetilde{z}_{ik}\Bigg)\end{aligned}$$

and, as $\gamma_{k\ell} = \frac{x_{k\ell}^{\tilde{\mathbf{z}}'\tilde{\mathbf{w}}}}{x_{k.}^{\tilde{\mathbf{z}}'} x_{.\ell}^{\tilde{\mathbf{w}}}}$ where $\tilde{\mathbf{z}}'$ is the previous probabilities, we have $\sum_\ell x_{i.} x_{.\ell}^{\tilde{\mathbf{w}}} \gamma_{k\ell} = \sum_\ell x_{i.} x_{.\ell}^{\tilde{\mathbf{w}}} \frac{x_{k\ell}^{\tilde{\mathbf{z}}'\tilde{\mathbf{w}}}}{x_{k.}^{\tilde{\mathbf{z}}'} x_{.\ell}^{\tilde{\mathbf{w}}}} = x_{i.} \frac{\sum_\ell x_{k\ell}^{\tilde{\mathbf{z}}'\tilde{\mathbf{w}}}}{x_{k.}^{\tilde{\mathbf{z}}}} = x_{i.}$ and therefore $\tilde{z}_{ik} = \arg\max_{\tilde{z}_{ik}} \sum_k \left(\tilde{z}_{ik} \log \pi_k + \tilde{z}_{ik} \sum_\ell x_{i\ell}^{\tilde{\mathbf{w}}} \log \gamma_{k\ell} - \tilde{z}_{ik} \log \tilde{z}_{ik}\right)$. The constraint $\sum_k \tilde{z}_{ik} = 1$ yields to $\tilde{z}_{ik} \propto \pi_k \exp(\sum_\ell x_{i\ell}^{\tilde{\mathbf{w}}} \log \gamma_{k\ell})$.

– Computation of $\tilde{\mathbf{w}}$: similarly, we have $\tilde{\mathrm{w}}_{j\ell} \propto \rho_\ell \exp(\sum_k x_{kj}^{\tilde{\mathbf{z}}} \log \gamma_{k\ell})$.

Finally, we obtain algorithm 4.3. In a similar way, the maximization of the $\mathrm{L_C}$ criterion can be performed by algorithm 4.4.

---

**Algorithm 4.3** Poisson LBVEM

---

**input:** x, $g$, $m$
**initialization:** $\tilde{\mathbf{z}}$, $\tilde{\mathbf{w}}$, $\pi_k = \frac{\tilde{z}_{.k}}{n}$, $\rho_\ell = \frac{\tilde{\mathrm{w}}_{.\ell}}{d}$, $\gamma_{k\ell} = \frac{x_{k\ell}^{\tilde{\mathbf{z}}\tilde{\mathbf{w}}}}{x_{k.}^{\tilde{\mathbf{z}}} x_{.\ell}^{\tilde{\mathbf{w}}}}$
**repeat**
 $x_{i\ell}^{\tilde{\mathbf{w}}} = \sum_j \tilde{\mathrm{w}}_{j\ell} x_{ij}$
 **repeat**
  **step 1.** $\tilde{z}_{ik} \propto \pi_k \exp(\sum_\ell x_{i\ell}^{\tilde{\mathbf{w}}} \log \gamma_{k\ell})$
  **step 2.** $\pi_k = \frac{\tilde{z}_{.k}}{n}$, $\gamma_{k\ell} = \frac{\sum_i \tilde{z}_{ik} x_{i\ell}^{\tilde{\mathbf{w}}}}{\sum_i (\tilde{z}_{ik} x_{i.}) x_{.\ell}^{\tilde{\mathbf{w}}}}$
 **until** convergence
 $x_{kj}^{\tilde{\mathbf{z}}} = \sum_i \tilde{z}_{ik} x_{ij}$
 **repeat**
  **step 3.** $\tilde{\mathrm{w}}_{j\ell} \propto \rho_\ell \exp(\sum_k x_{kj}^{\tilde{\mathbf{z}}} \log \gamma_{k\ell})$
  **step 3.** $\rho_\ell = \frac{\tilde{\mathrm{w}}_{.\ell}}{d}$, $\gamma_{k\ell} = \frac{\sum_j \tilde{\mathrm{w}}_{j\ell} x_{kj}^{\tilde{\mathbf{z}}}}{\sum_j (\tilde{\mathrm{w}}_{j\ell} x_{.j}) x_{k.}^{\tilde{\mathbf{z}}}}$
 **until** convergence
**until** convergence
**return** $\pi$, $\rho$, $\gamma$

---

REMARK 4.1.– When the proportions are assumed to be equal ($\pi_1 = \ldots = \pi_g$ and $\rho_1 = \ldots = \rho_m$), it can be observed that the Poisson LBCEM algorithm is equivalent to the CROINFO algorithm. Moreover, after the step of computation of $\gamma$, the complete data log-likelihood $\mathrm{L_C}(\mathbf{z}, \mathbf{w}, \gamma(\mathbf{z}, \mathbf{w}))$ becomes

$$
\begin{aligned}
\mathrm{L_C}(\mathbf{z}, \mathbf{w}, \gamma(\mathbf{z}, \mathbf{w})) &= \sum_{i,j,k,\ell} z_{ik} w_{j\ell} (x_{ij} \log \gamma_{k\ell} - x_{i.} x_{.j} \gamma_{k\ell}) \\
&= \sum_{k,\ell} x_{k\ell}^{\mathbf{zw}} \log \frac{x_{k\ell}^{\mathbf{zw}}}{x_{k.}^{\mathbf{z}} x_{.\ell}^{\mathbf{w}}} - N \\
&= N \sum_{k,\ell} p_{k\ell}^{\mathbf{zw}} \log \frac{p_{k\ell}^{\mathbf{zw}}}{p_{k.}^{\mathbf{z}} p_{.\ell}^{\mathbf{w}}} + N(\log N - 1) \\
&= N \mathcal{I}(P_{KL}^{\mathbf{zw}}) + C
\end{aligned}
$$

where $C$ is a constant and therefore the maximization of the complete data log-likelihood of the Poisson LBM with equal proportions is equivalent to the maximization of the information criterion $\mathcal{I}(P_{KL}^{\mathbf{zw}})$ which shows that the maximization of $\mathcal{I}$ assumes implicitly that the proportions of rows and rows are equal. Hence, considering the unknown proportions, Poisson LBCEM clearly appears as an extension of CROINFO. The assignation and maximization are equivalent. This assumption (equality of proportions) usually not true in practice can lead to bad results.

## 4.5. Comparison of all algorithms

In this chapter, two unified frameworks have been presented: in the first framework, based on measures of association, the phi-squared coefficient and mutual information have been retained. In the second framework, using the model-based co-clustering approach, the block model and Poisson LBM have been considered. Both approaches yielded four algorithms, CROKI2, CROINFO, Poisson LBVEM and LBCEM summarized in Table 4.8.

**Algorithm 4.4** Poisson LBCEM

**input:** x, $g$, $m$

**initialization:** $\widetilde{z}$, $\widetilde{\mathrm{w}}$, $\pi_k = \frac{z_{.k}}{n}$, $\rho_\ell = \frac{w_{.\ell}}{d}$, $\gamma_{k\ell} = \frac{x^{bz\mathbf{w}}_{k\ell}}{x^{\mathbf{z}}_{k.}x^{\mathbf{w}}_{.\ell}}$

**repeat**

  $x^{\mathbf{w}}_{i\ell} = \sum_j w_{j\ell}x_{ij}$

  **repeat**

    **step 1.** $z_i = \arg\max_k \left(\log \pi_k + \sum_\ell x^{\widetilde{\mathbf{w}}}_{i\ell} \log \gamma_{k\ell}\right)$

    **step 2.** $\pi_k = \frac{z_{.k}}{n}$, $\gamma_{k\ell} = \frac{\sum_i z_{ik}x^{\mathbf{w}}_{i\ell}}{\sum_i (z_{ik}x_{i.})x^{\mathbf{w}}_{.\ell}}$

  **until** convergence

  **repeat**

    $x^{\mathbf{z}}_{kj} = \sum_j z_{ik}x_{ij}$

    **step 3.** $w_j = \arg\max_\ell \left(\log \rho_\ell + \sum_k x^{\widetilde{\mathbf{z}}}_{kj} \log \gamma_{k\ell}\right)$

    **step 3.** $\rho_\ell = \frac{\widetilde{\mathrm{w}}_{.\ell}}{d}$, $\gamma_{k\ell} = \frac{\sum_j w_{j\ell}x^{\mathbf{z}}_{kj}}{\sum_j (w_{j\ell}x_{.j})x^{\mathbf{z}}_{k.}}$

  **until** convergence

**until** convergence

**return** z, w, $\pi$, $\rho$, $\gamma$

| Algorithm | Criterion |
|---|---|
| CROKI2 | $\widetilde{W}_{\Phi^2}(\mathbf{z},\mathbf{w},\boldsymbol{\delta}) = \sum_{i,j,k,\ell} z_{ik}w_{j\ell}\, p_{i.}p_{.j} \left(\frac{p_{ij}}{p_{i.}p_{.j}} - \delta_{k\ell}\right)^2$ |
| CROINFO | $\widetilde{W}_{\mathcal{I}}(\mathbf{z},\mathbf{w},\boldsymbol{\delta}) - \sum_{i,j} p_{ij} \log \frac{p_{ij}}{p_{i.}p_{.j}} - \sum_{k,\ell} p^{\mathbf{zw}}_{k\ell} \log \delta_{k\ell}$ |
| PLBCEM | $L_C(\mathbf{z},\mathbf{w};\boldsymbol{\theta}) = \sum_{i,j,k,\ell} z_{ik}w_{j\ell}(x_{ij}\log\gamma_{k\ell} - x_{i.}x_{.j}\gamma_{k\ell})$ $+ \sum_{i,k} z_{ik}\log\pi_k + \sum_{j,\ell} w_{j\ell}\log\rho_\ell$ |
| PLBVEM | $F_C(\tilde{\mathbf{z}},\tilde{\mathbf{w}};\boldsymbol{\theta}) = \sum_{i,j,k,\ell} \tilde{z}_{ik}\tilde{w}_{j\ell}(x_{ij}\log\gamma_{k\ell} - x_{i.}x_{.j}\gamma_{k\ell})$ $+\sum_{i,k} \tilde{z}_{ik}\log\pi_k + \sum_{j,\ell} \tilde{w}_{j\ell}\log\rho_\ell$ $-\sum_{i,k}\tilde{z}_{ik}\log\tilde{z}_{ik} - \sum_{j,\ell}\tilde{w}_{j\ell}\log\tilde{w}_{j\ell}$ |

**Table 4.8.** *Algorithms and criteria*

In this section, we present case studies to demonstrate the difference among these algorithms. First, we compare CROKI2 and CROINFO in terms of convergence. Second, we show the impact of the proportions on the co-clustering performance of CROINFO. Third, we compare Poisson LBCEM and LBVEM in terms of co-clustering and estimation. Finally, we compare all the algorithms discussed in this chapter.

### 4.5.1. CROKI2 *versus* CROINFO

First, we study the behavior of the different criteria $\Phi^2$ and $\mathcal{I}$ when applying CROKI2 and CROINFO. As co-clustering has been a topic of much interest in document clustering, we propose to use the Classic3 data set, a document collection from the SMART project at Cornell University. This data set consists of 3,891 documents and 4,303 words after removing stop words. It is classified into three classes denoted as *Medline* (1,033 documents), *Cisi* (1,460 documents) and *Cranfield* (1,398 documents). Note that the proportions of clusters are not dramatically different. To visualize this data set, we use the CA and we present, in Figure 4.5, the projection of rows and columns into the factorial plane spawned by the first and second axes.

We applied the two algorithms with $g = 3$ and $m = 3$. The CROKI2 algorithm naturally increases the objective function $\Phi^2$ and the CROINFO algorithm increases $\mathcal{I}$. However, in our experiments, we have observed that each algorithm monotonically increases the two objective functions. We illustrate this behavior in Figure 4.6.

### 4.5.2. CROINFO *versus Poisson* LBCEM

As we discussed in section 4.4.3, the CROINFO and PLBCEM algorithms are the same when the proportions are assumed to be equal. We simulate a data set with dramatically different proportions ($\pi_1 = 0.1$, $\pi_2 = 0.9$, $\rho_1 = 0.178$ and $\rho_2 = 0.822$). We observe the projection of the sets of rows and columns by CA in Figure 4.7. In Tables 4.9 and 4.10, we present the confusion matrices between estimated $(\hat{\mathbf{z}}, \hat{\mathbf{w}})$, obtained by the application of both algorithms with $g = 2$ and $m = 2$, and the true partitions $(\mathbf{z}^T, \mathbf{w}^T)$. We note the good performance of Poisson LBCEM and therefore the impact of absence of proportions in the CROINFO criteria.

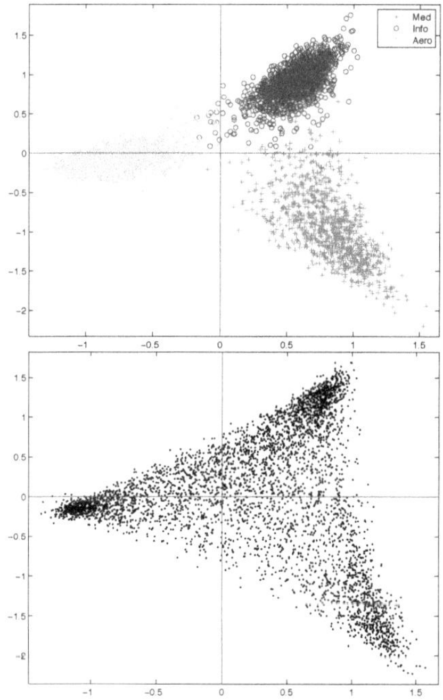


**Figure 4.5.** *a) Projection of the rows and b) the columns into the factorial plane spawned by the first and second axes*

| | $\mathbf{z}_1^T$ | $\mathbf{z}_2^T$ |
|---|---|---|
| $\hat{\mathbf{z}}_1$ | 109 | 1 |
| $\hat{\mathbf{z}}_2$ | 213 | 677 |

| | $\mathbf{w}_1^T$ | $\mathbf{w}_2^T$ |
|---|---|---|
| $\hat{\mathbf{w}}_1$ | 336 | 75 |
| $\hat{\mathbf{w}}_2$ | 1 | 88 |

**Table 4.9.** *Confusion matrices between estimated partitions* $(\hat{\mathbf{z}}, \hat{\mathbf{w}})$ *obtained by* CROINFO *and true partitions*$(\mathbf{z}^T, \mathbf{w}^T)$

| | $\mathbf{z}_1^T$ | $\mathbf{z}_2^T$ |
|---|---|---|
| $\hat{\mathbf{z}}_1$ | 103 | 7 |
| $\hat{\mathbf{z}}_2$ | 1 | 889 |

| | $\mathbf{w}_1^T$ | $\mathbf{w}_2^T$ |
|---|---|---|
| $\hat{\mathbf{w}}_1$ | 411 | 0 |
| $\hat{\mathbf{w}}_2$ | 2 | 87 |

**Table 4.10.** *Confusion matrices between estimated* $(\hat{\mathbf{z}}, \hat{\mathbf{w}})$ *obtained by Poisson* LBCEM *and true* $(\mathbf{z}^T, \mathbf{w}^T)$

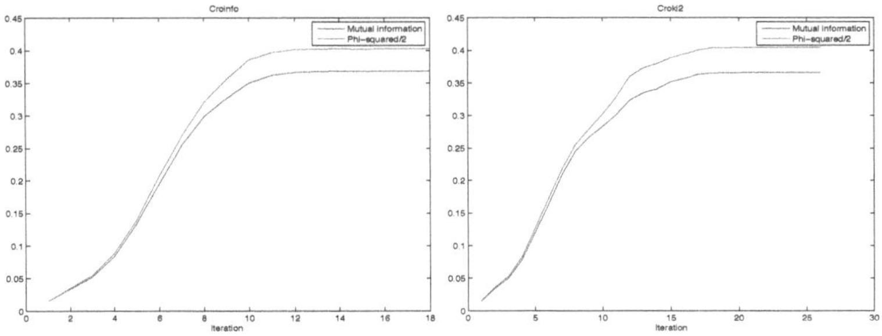


**Figure 4.6.** *Behavior of mutual information and phi-squared coefficient as a function of iterations by* CROINFO *and* CROKI2 *on Classic3 data*

### 4.5.3. *Poisson* LBVEM *versus Poisson* LBCEM

In this section, to illustrate the difference between the LBVEM and LBCEM algorithms in terms of behavior, we studied their performances on simulated data. In our experiments, we selected 12 types of data arising from $3 \times 2$ component mixture model corresponding to three degrees of cluster overlap (well separated, moderately separated and ill-separated), and four data dimensions ($n \times d = 50 \times 50$, $n \times d = 100 \times 60, n \times d = 200 \times 120$ and $n \times d = 300 \times 180$).

The concept of cluster separation is difficult to visualize for Poisson LBMs, but the degree of overlap can be measured by the true error rate, which is defined as the average misclassification probability $\mathbf{E}(\Gamma((\mathbf{z}, \mathbf{w}), r_B(\mathbf{x})))$ where $r_B$ is the optimal Bayes' rule $r_B(\mathbf{x}) = \arg\max_{\mathbf{z},\mathbf{w}} P(\mathbf{z}, \mathbf{w}|\mathbf{x})$ associated with the LBM and $\Gamma((\mathbf{z}, \mathbf{w}), (\mathbf{z}^T, \mathbf{w}^T))$ is the proportion of misclassified items. Its computation being theoretically difficult, we used Monte Carlo simulations and approximated this error rate by comparing the partitions simulated with those we obtained by applying a classification step. Parameters were selected so as to obtain error rates, respectively, in [0.04, 0.06], for the well-separated (+), in [0.14, 0.16], for the moderately (++), and in [0.23, 0.26], for the ill-separated situations (+++).

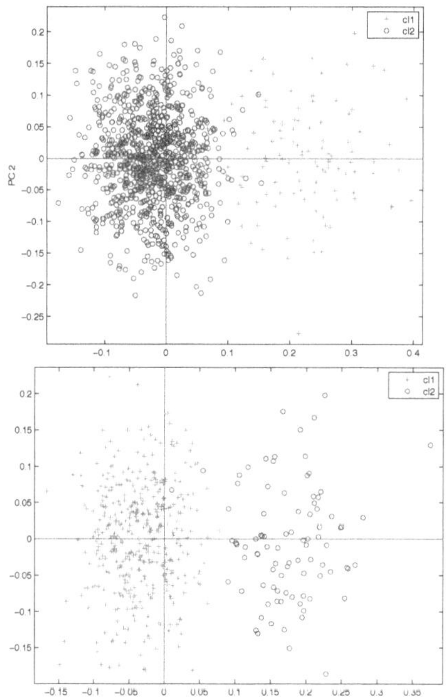


**Figure 4.7.** *a) Row representation onto plane 1,2 and b) column representation onto plane 1,2*

For each of these 12 data structures, 30 samples were generated and, for each sample, Poisson LBVEM and Poisson LBCEM algorithms were run 20 times starting from the same random situations. Then, the best solution for each method was selected. The simulation results are summarized in Tables 4.11 and 4.12. The first one displays the mean error rates and the second displays the mean Euclidean distance between true parameters and estimated parameters for each situation.

| Degree of overlap | Size | PLBCEM | PLBVEM |
|---|---|---|---|
| + | 50 × 50 | 0.0690 | 0.0709 |
| | 100 × 60 | 0.0647 | 0.0594 |
| | 200 × 120 | 0.0554 | 0.0539 |
| | 300 × 180 | 0.0576 | 0.0561 |
| ++ | 50 × 50 | 0.2740 | 0.2900 |
| | 100 × 60 | 0.1870 | 0.1768 |
| | 200 × 120 | 0.1617 | 0.1533 |
| | 300 × 180 | 0.1459 | 0.1372 |
| +++ | 50 × 50 | 0.3418 | 0.3467 |
| | 100 × 60 | 0.2895 | 0.2937 |
| | 200 × 120 | 0.2488 | 0.2262 |
| | 300 × 180 | 0.2635 | 0.2341 |

**Table 4.11.** *Comparison of Poisson* LBVEM *and Poisson* LBCEM *in terms of mean error rates between estimated* $(\mathbf{z}, \mathbf{w})$ *and true partitions* $(\mathbf{z}^T, \mathbf{w}^T)$

| Degree of overlap | Size | PLBCEM | PLBVEM |
|---|---|---|---|
| + | 50 × 50 | 0.0061 | 0.0060 |
| | 100 × 60 | 0.0032 | 0.0030 |
| | 200 × 120 | 0.0015 | 0.0013 |
| | 300 × 180 | 0.0012 | 0.0011 |
| ++ | 50 × 50 | 0.0195 | 0.0178 |
| | 100 × 60 | 0.0048 | 0.0039 |
| | 200 × 120 | 0.0024 | 0.0018 |
| | 300 × 180 | 0.0013 | 0.0011 |
| +++ | 50 × 50 | 0.0131 | 0.0109 |
| | 100 × 60 | 0.0068 | 0.0062 |
| | 200 × 120 | 0.0027 | 0.0019 |
| | 300 × 180 | 0.0024 | 0.0015 |

**Table 4.12.** *Comparison of Poisson* LBVEM *and Poisson* LBCEM *in terms of mean Euclidean distance between true parameters and estimated parameters*

The main findings arising from these experiments are the following. In terms of co-clustering, the Poisson LBVEM algorithm, even though it is slower than LBCEM, generally gives the best results, especially when the clusters are not well separated; not surprisingly, its performance increases with the size of the data. In terms of quality of estimation, LBVEM outperforms LBCEM (Table 4.12). This quality increases with the size of the data. This point provides an additional argument for its use in the selection model context.

### 4.5.4. *Behavior of* CROKI2, CROINFO, LBCEM *and* LBVEM

The four algorithms LBCEM, LBVEM, CROINFO and CROKI2 were evaluated on Classic3 data set in varying the number of column clusters $m$. The values of $\Phi^2$, $\mathcal{I}$ and the number of documents misclassified denoted as $e$ are presented in Table 4.13.

– We observe that the approximation of $\mathcal{I}$ by $\frac{1}{2}\Phi^2$ is very good when $m$ is high enough, which is consistent with proposition 4.1.

– Unsurprisingly, on the one hand, CROKI2 maximizes $\Phi^2$ and, on the other hand, CROINFO maximizes $\mathcal{I}$. We note that in all situations, CROINFO outperforms CROKI2 in terms of clustering. The mutual information appears more adapted to co-clustering.

– We observe that implicit dimensionality reduction by co-clustering actually gives better document clusters, in the sense that the cluster labels agree more with the true class labels with fewer word clusters. However, we note that the values of $\Phi^2$ and $\mathcal{I}$ increase with $m$ even though this growth becomes smaller from $m = 30$. The quality of clustering by CROKI2 and CROINFO starts to decrease from $m = 100$.

| $m$ | Algorithm | $\Phi^2$ | $\mathcal{I}$ | $e$ |
|---|---|---|---|---|
| | CROINFO | 0.8047575 | 0.3682842 | 52 |
| 3 | CROKI2 | 0.8094602 | 0.3645942 | 64 |
| | PLBCEM | 0.8044226 | 0.3682617 | 52 |
| | PLBVEM | 0.8036757 | 0.3678993 | 52 |
| | CROINFO | 0.9248306 | 0.4522520 | 31 |
| 5 | CROKI2 | 0.9393426 | 0.4452273 | 51 |
| | PLBCEM | 0.9240040 | 0.4522505 | 32 |
| | PLBVEM | 0.9225218 | 0.4513789 | 28 |
| | CROINFO | 0.9862210 | 0.4990168 | 29 |
| 10 | CROKI2 | 0.9973430 | 0.4924405 | 40 |
| | PLBCEM | 0.9844507 | 0.4988628 | 28 |
| | PLBVEM | 0.9838697 | 0.4965514 | 29 |
| | CROINFO | 1.0193666 | 0.5270878 | 28 |
| 30 | CROKI2 | 1.0233885 | 0.5204656 | 47 |
| | PLBCEM | 1.0095892 | 0.5195290 | 26 |
| | PLBVEM | 1.0095892 | 0.5195290 | 26 |
| | CROINFO | 1.0226414 | 0.5302415 | 32 |
| 40 | CROKI2 | 1.0266164 | 0.5257123 | 36 |
| | PLBCEM | 1.0124342 | 0.5220182 | 25 |
| | PLBVEM | 1.0124342 | 0.5220182 | 25 |
| | CROINFO | 1.0240256 | 0.5320448 | 28 |
| 50 | CROKI2 | 1.0282436 | 0.5282652 | 37 |
| | PLBCEM | 1.0142280 | 0.5232399 | 28 |
| | PLBVEM | 1.0142280 | 0.5232399 | 28 |
| | CROINFO | 1.0289786 | 0.5350597 | 30 |
| 100 | CROKI2 | 1.0310344 | 0.5319551 | 48 |
| | PLBCEM | 1.0159995 | 0.5255252 | 26 |
| | PLBVEM | 1.0159995 | 0.5255252 | 26 |

**Table 4.13.** *Performances of the four algorithms on Classic3 data set; $e$ denotes the number of documents misclassified*

– Poisson LBCEM which is an extension of CROINFO appears more efficient, although the proportions are not dramatically different. We have discussed the impact of these parameters in section 4.5.2.

– As should be expected, the values of $\mathcal{I}$ obtained by LBCEM are smaller than those obtained by CROINFO but in all cases, the values obtained by these algorithms are very close. This is due to the relation between the criteria optimized by Poisson LBCEM and CROINFO.

– In all situations, LBVEM outperforms all the other algorithms. This performance was observed for other data sets and already confirmed by intensive numerical experiments on binary data (see, for instance, [GOV 05, GOV 08]).

## 4.6. Conclusion

In this chapter, two unified frameworks have been studied. In the first framework, based on measures of association, we have retained phi-squared coefficient and mutual information criteria. In the second framework, based on model-based co-clustering, we have considered the block model and Poisson LBM. Both approaches yielded four algorithms: CROKI2 CROINFO and Poisson LBVEM and Poisson LBCEM . We have established important connections between these algorithms. Note that our two approaches could be extended to other measures of association and other LBMs. In terms of performance, we have noted that Poisson LBVEM outperforms all the other algorithms. We have also selected some situations showing the importance of proportions in co-clustering and therefore the weakness of algorithms omitting this situation such as CROKI2 and CROINFO .

In an objective visualization, combining clustering and reduction for mapping clusters rather than individuals is therefore an appealing requirement for data analysis. One way to perform this task is by showing the clusters on a map after performing an *ad hoc* algorithm for partitioning the data set and reducing the data space, both separately as we have performed by using co-clustering and CA. Alternatively, Kohonen's self-organizing map (SOM) [KOH 82] enables clustering and mapping of clusters while providing one final single map. The SOM algorithm is not derived exactly through the optimization of an objective function, and some crucial parameters need to be chosen empirically. A probabilistic model for SOM is appealing for several reasons, the principal reason being that a parametric model is flexible and scalable when defined appropriately. Generative topographic mapping (GTM) [BIS 98] is a parametric SOM that offers a number of advantages compared to the standard SOM. Recently, in [PRI 12], the authors presented a GTM based on the Poisson latent block mixture model. The empirical results obtained showed that the method proposed is able to present a quick summary of the data set contents. In addition, the proposed model is parsimonious when compared to the existing alternative in the domain.

Chapter 5

# Co-Clustering of Continuous Data

In this chapter, we focus on the co-clustering of an $n \times d$ data matrix $\mathbf{x} = (x_{ij})$ corresponding to the observations of a set $J$ of $d$ continuous variables on a set $I$ of $n$ individuals. As for the principal component analysis (PCA), we associate with the set of individuals weights $\mathbf{p} = (p_1, \ldots, p_n)$ and with the set of variables weights $\mathbf{q} = (q_1, \ldots, q_d)$. Moreover, we consider the data matrix $\mathbf{x}$ to be "column-centered", i.e. the means of the columns of $\mathbf{x}$ are equal to 0. Note that, if it is not true, it is easy to modify the initial table to obtain this property. Finally, the initial data that are defined by a "column-centered" matrix $\mathbf{x}$ and two vectors $\mathbf{p}$ and $\mathbf{q}$ will be represented by the triple $(\mathbf{x}, \mathbf{p}, \mathbf{q})$.

To analyze this type of data, we can use PCA that summarizes data by means of new axes. In this chapter, the aim is to summarize the data by means of clusters of the individuals and the variables. There are different ways to attempt this block clustering objective. The simple one consists of clustering the set of variables and, using the data obtained by replacing each variable with the mean of its cluster, applying a clustering algorithm to the set of individuals. For cluster variables, we can use partitioning methods such as $k$-means, or hierarchical methods such as the well-known SAS Varclus procedure, which is a type of oblique component analysis related to multiple group factor analysis [HAR 76]. The $k$-means applied on centered variables and Varclus are used as variable-reduction methods where the variables of the same cluster are correlated as much as possible, and two variables belonging to different clusters are as uncorrelated as possible.

In order to obtain homogeneous blocks, there are more adapted methods that consist of clustering both sets *simultaneously*. Most methods used in this context can be grouped into two families: metric methods and model-based methods. In the first family, which aims to optimize an objective function, we can cite, for instance, the works of Hartigan [HAR 72], Bock [BOC 79] and Govaert [GOV 83, GOV 95]. In the second family, two approaches can be considered: the latent block model with Gaussian distributions and the mixture model with conditional independent Gaussian distributions more adapted to continuous data.

This chapter is organized as follows. Section 5.1 describes, the metric methods, which can be viewed as a minimization of a loss information function. We describe the CROEUC algorithm used for this purpose and, before concluding, we present an example. In section 5.2, we study the latent block model with Gaussian distributions and associated algorithms Gaussian LBCEM and LBVEM that we illustrate by an example in section 5.3. In section 5.4, we present a Gaussian block mixture model and derive two algorithms that we evaluate in section 5.5.

## 5.1. Metric approach

Two geometrical representations can be associated with continuous data:

– A geometrical representation of the individuals by a set of $n$ points of $\mathbb{R}^d$ where the coordinates of the $n$ points are the rows of the data matrix x, the $p_i$ are the weights of the point and the $q_j$ can be used to define the Euclidean metric $\mathrm{d}^2(i,i') = \sum_j q_j(x_{ij} - x_{i'j})^2$.

– A geometrical representation of the variables by a set of $d$ points of $\mathbb{R}^n$ where the coordinates of the $n$ points are the columns of the data matrix x, the $q_j$ are the weights of the point and the $p_i$ can be used to define the Euclidean metric $\mathrm{d}^2(j,j') = \sum_i p_i(x_{ij} - x_{ij'})^2$.

In the following, and only to simplify the notation, we assume that $p_i = \frac{1}{n}$ for all $i$ and $q_j = 1$ for all $j$.

### 5.1.1. *Measure of information*

The information measure we would like to preserve is the following:

$$\mathcal{I}(\mathbf{x}, \mathbf{p}, \mathbf{q}) = \sum_{i,j} p_i q_j x_{ij}^2 = \frac{1}{n} \sum_{i,j} x_{ij}^2.$$

With the geometrical representations, this information represents in $\mathbb{R}^d$ the inertia of the set $I$ relative to the center of gravity and in $\mathbb{R}^n$ the inertia of the set $J$ relative to the origin. Let us note that this information measure is the measure used by PCA.

### 5.1.2. *Summarized data associated with partitions*

As was seen in section 2.1 of Chapter 2, the continuous data matrix can be summarized by a smaller continuous matrix by associating each block with its mean (see Figure 5.1). More precisely, using a partition z of $I$ and a partition w of $J$, the initial data are summarized by two sets of weights $\mathbf{p}^{\mathbf{z}} = (p_1^{\mathbf{z}}, \ldots, p_g^{\mathbf{z}})$ and $\mathbf{q}^{\mathbf{w}} = (q_1^{\mathbf{w}}, \ldots, q_m^{\mathbf{w}})$ and a $g \times m$ matrix $\mathbf{x}^{\mathbf{zw}} = (x_{k\ell}^{\mathbf{zw}})$ defined by

$$p_k^{\mathbf{z}} = \frac{\sum_i z_{ik}}{n} = \frac{z_{.k}}{n}, \qquad q_\ell^{\mathbf{w}} = \sum_j w_{j\ell} = w_{.\ell}$$

and

$$x_{k\ell}^{\mathbf{zw}} = \frac{\sum_{i,j} z_{ik} w_{j\ell} p_i q_j x_{ij}}{\sum_{i,j} z_{ik} w_{j\ell} p_i q_j} = \frac{\sum_{i,j} z_{ik} w_{j\ell} x_{ij}}{z_{.k} w_{.\ell}}.$$

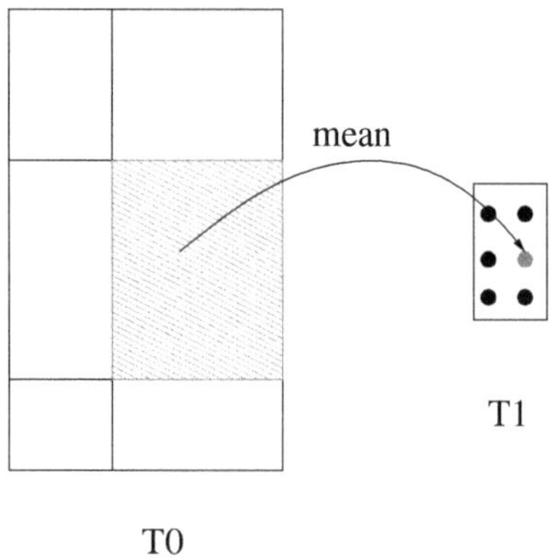


**Figure 5.1.** *Aggregation of data matrix in summary matrix using co-clustering*

We have, thus, associated the two partitions $\mathbf{z}$ and $\mathbf{w}$ with a new triple $(\mathbf{x}^{\mathbf{zw}}, \mathbf{p}^{\mathbf{z}}, \mathbf{q}^{\mathbf{w}})$ which has the same structure as the initial data $(\mathbf{x}, \mathbf{p}, \mathbf{q})$ and therefore we can define the information measure

$$\mathcal{I}(\mathbf{x}^{\mathbf{zw}}, \mathbf{p}^{\mathbf{z}}, \mathbf{q}^{\mathbf{w}}) = \sum_{k,\ell} p_k^{\mathbf{z}} q_\ell^{\mathbf{w}} (x_{k\ell}^{\mathbf{zw}})^2 = \frac{1}{n} \sum_{k,\ell} z_{.k} w_{.\ell} (x_{k\ell}^{\mathbf{zw}})^2. \qquad [5.1]$$

We have illustrated this summary matrix in Table 5.1. Starting from the data $\mathbf{x}$ with weights $p_i = 1/4$, $q_j = 1$ and partitions $\mathbf{z} = (1, 1, 2, 2)$ and $\mathbf{w} = (1, 1, 2)$, we obtain the summary $\mathbf{x}^{\mathbf{zw}}$ with weights $\mathbf{p}^{\mathbf{z}} = (1/2, 1/2)$ and $\mathbf{q}^{\mathbf{w}} = (2, 1)$.

$$\mathbf{x} = \begin{pmatrix} 1 & 2 & 8 \\ 2 & 1 & 7 \\ 2 & 4 & 7 \\ 4 & 4 & 6 \end{pmatrix} \qquad \mathbf{x}^{\mathbf{zw}} = \begin{pmatrix} 1.5 & 7.5 \\ 3.5 & 6.5 \end{pmatrix}$$

**Table 5.1.** *Example of summary*

As in Chapters 3 and 4, intermediate matrices $\mathbf{x}^{\mathbf{w}} = (x_{i\ell}^{\mathbf{w}})$ of size $(n \times m)$ and $\mathbf{x}^{\mathbf{z}} = (x_{kj}^{\mathbf{z}})$ of size $(g \times m)$ can be defined but this time we have

$$x_{i\ell}^{\mathbf{w}} = \frac{\sum_{j,\ell} w_{j\ell} q_j x_{ij}}{\sum_{j,\ell} w_{j\ell} q_j} = \frac{\sum_{j,\ell} w_{j\ell} x_{ij}}{w_{.\ell}} \quad \text{and}$$

$$x_{kj}^{\mathbf{z}} = \frac{\sum_{i,k} z_{ik} p_i x_{ij}}{\sum_{i,k} p_i z_{ik}} = \frac{\sum_{i,k} z_{ik} x_{ij}}{z_{.k}}$$

and with the data defined in Table 5.1, these intermediate matrices become

$$\mathbf{x}^{\mathbf{z}} = \begin{pmatrix} 1.5 & 1.5 & 7.5 \\ 3 & 4 & 6.5 \end{pmatrix} \quad \text{and} \quad \mathbf{x}^{\mathbf{w}} = \begin{pmatrix} 1.5 & 8 \\ 1.5 & 7 \\ 3 & 7 \\ 4 & 6 \end{pmatrix}.$$

In the following, we will denote $(\mathbf{x}^{\mathbf{w}}, \mathbf{p}, \mathbf{q}^{\mathbf{w}})$ as the triple obtained when z is the singleton partition and $(\mathbf{x}^{\mathbf{z}}, \mathbf{p}^{\mathbf{z}}, \mathbf{q})$ as the triple obtained when w is the singleton partition. Hence, we obtain the associated measures of association as follows

$$\mathcal{I}(\mathbf{x}^{\mathbf{z}}, \mathbf{p}^{\mathbf{z}}, \mathbf{q}) = \frac{1}{n} \sum_{k,j} z_{.k} (x_{kj}^{\mathbf{z}})^2 \text{ and } \mathcal{I}(\mathbf{x}^{\mathbf{w}}, \mathbf{p}, \mathbf{q}^{\mathbf{w}}) = \frac{1}{n} \sum_{i,\ell} w_{.\ell} (x_{i\ell}^{\mathbf{w}})^2.$$

REMARK 5.1.– When w is the partition of singletons, this criterion can be expressed as the loss of information due to the partition z and, by using the Huygens theorem, it can be shown that

$$\mathcal{I}(\mathbf{x}, \mathbf{p}, \mathbf{q}) - \mathcal{I}(\mathbf{x}^{\mathbf{z}}, \mathbf{p}^{\mathbf{z}}, \mathbf{q}) = \frac{1}{n} \mathrm{W}(\mathbf{z}|J)$$

where $\mathrm{W}(\mathbf{z}|J) = \sum_{i,k} z_{ik} \sum_j (x_{ij} - x^{\mathbf{z}}_{kj})^2$ is the intra class inertia, or within-group sum of squares, minimized by the classical $k$-means algorithm. Similarly, when z is the partition of singletons, we have

$$\mathcal{I}(\mathbf{x}, \mathbf{p}, \mathbf{q}) - \mathcal{I}(\mathbf{x}^{\mathbf{w}}, \mathbf{p}, \mathbf{q}^{\mathbf{w}}) = \frac{1}{n}\mathrm{W}(\mathbf{w}|I)$$

where $\mathrm{W}(\mathbf{w}|I) = \sum_{j,\ell} w_{j\ell} \sum_i (x_{ij} - x^{\mathbf{w}}_{i\ell})^2$.

### 5.1.3. *Objective function*

It can be easily proved that the measure of information associated with the summarized data $(\mathbf{x}^{\mathbf{zw}}, \mathbf{p}^{\mathbf{z}}, \mathbf{q}^{\mathbf{w}})$ is smaller than the information associated with the initial data, i.e. grouping rows and columns leads to a loss of information. Therefore, the objective will be to minimize the objective function

$$\begin{aligned} \mathrm{W}(\mathbf{z}, \mathbf{w}) &= \mathcal{I}(\mathbf{x}, \mathbf{p}, \mathbf{q}) - \mathcal{I}(\mathbf{x}^{\mathbf{zw}}, \mathbf{p}^{\mathbf{z}}, \mathbf{q}^{\mathbf{w}}) \\ &= \frac{1}{n} \sum_{i,j,k,\ell} z_{ik} w_{j\ell} (x_{ij} - x^{\mathbf{zw}}_{k\ell})^2 \end{aligned} \qquad [5.2]$$

or, equivalently, to maximize the measure of information $\mathcal{I}(\mathbf{x}^{\mathbf{zw}}, \mathbf{p}^{\mathbf{z}}, \mathbf{q}^{\mathbf{w}})$.

Note that the objective function defined in equation [5.2] can also be written as

$$||\mathbf{x} - \mathbf{y}||^2 = \mathrm{trace}\left((\mathbf{x} - \mathbf{y})(\mathbf{x} - \mathbf{y})^t\right), \qquad [5.3]$$

where $\mathbf{y} = \mathbf{z}(\mathbf{x}^{\mathbf{zw}})\mathbf{w}^t$ can be viewed as a reconstructed matrix based on the cluster structures. For instance, in the previous example, the matrix y is

$$\begin{pmatrix} 1.5 & 1.5 & 7.5 \\ 1.5 & 1.5 & 7.5 \\ 3.5 & 3.5 & 6.5 \\ 3.5 & 3.5 & 6.5 \end{pmatrix}.$$

The clustering problem can therefore be formulated as a matrix approximation problem where the clustering aim is to minimize the approximation error between the original data x and the reconstructed matrix y based on the cluster structures. This approximation can be solved by an iterative alternating least-squares optimization procedure (see, for instance, [BAI 97, VIC 01, CHO 04, LI 05, ROS 09, LAB 11a]). These algorithms are equivalent and consist of using the principle of a double $k$-means. A version based on update rules is presented in algorithm 5.1. Furthermore, we recommend another version called CROEUC [GOV 83, GOV 95], which is based on the use of reduced intermediate matrices $\mathrm{x}^{\mathrm{w}}$ and $\mathrm{x}^{\mathrm{z}}$ and therefore requires less computation.

**Algorithm 5.1** Double $k$-means

**Input:** x, $g$, $m$
**Initialization:** z, w, $x_{k\ell}^{\mathbf{zw}} = \sum_{i,j} \frac{z_{ik}w_{j\ell}x_{ij}}{z_{.k}w_{.\ell}}$
**repeat**
  **step 1.** $z_i = \arg\min_k \sum_{j,\ell} w_{j\ell}(x_{ij} - x_{k\ell}^{\mathbf{zw}})^2$
  **step 2.** $w_j = \arg\min_\ell \sum_{i,k} z_{ik}(x_{ij} - x_{k\ell}^{\mathbf{zw}})^2$
  **step 3.** $x_{k\ell}^{\mathbf{zw}} = \sum_{i,j} \frac{z_{ik}w_{j\ell}x_{ij}}{z_{.k}w_{.\ell}}$
**until** convergence
**return** z, w

### 5.1.4. CROEUC *algorithm*

Using the strategy described in section 2.4.1 of Chapter 2, the CROEUC algorithm consists of computing, for a fixed partition of J, the best partition of $I$, and for a fixed partition $J$, the best partition of $I$. For this, it can be noted that following the Huygens theorem, we have

$$\mathcal{I}(\mathbf{x}^{\mathbf{w}}, \mathbf{p}, \mathbf{q}^{\mathbf{w}}) - \mathcal{I}(\mathbf{x}^{\mathbf{zw}}, \mathbf{p}^{\mathbf{z}}, \mathbf{q}^{\mathbf{w}}) = \mathrm{W}(\mathbf{z}|\mathbf{w}),$$

where

$$\mathrm{W}(\mathbf{z}|\mathbf{w}) = \sum_{i,k}\sum_{\ell} w_{.\ell}(x_{i\ell}^{\mathbf{w}} - x_{k\ell}^{\mathbf{zw}})^2$$

is the intra class inertia criterion for the partition z when the data are the triple $(\mathbf{x}^{\mathbf{w}}, \mathbf{p}, \mathbf{q}^{\mathbf{w}})$ and

$$\mathcal{I}(\mathbf{x}^{\mathbf{z}}, \mathbf{p}^{\mathbf{z}}, \mathbf{q}) - \mathcal{I}(\mathbf{x}^{\mathbf{zw}}, \mathbf{p}^{\mathbf{z}}, \mathbf{q}^{\mathbf{w}}) = \mathrm{W}(\mathbf{w}|\mathbf{z}),$$

where

$$\mathrm{W}(\mathbf{w}|\mathbf{z}) = \sum_{j,\ell} w_{j\ell} \sum_{k} z_{.k}(x^{\mathbf{z}}_{kj} - x^{\mathbf{zw}}_{k\ell})^2$$

is the intraclass inertia criterion for the partition w when the data are the triple $(\mathbf{x}^{\mathbf{z}}, \mathbf{p}^{\mathbf{z}}, \mathbf{q})$. Therefore, the minimization of the objective function [5.2] can be performed by alternating the $k$-means algorithm on the triple $(\mathbf{x}^{\mathbf{w}}, \mathbf{p}, \mathbf{q}^{\mathbf{w}})$ and on the triple $(\mathbf{x}^{\mathbf{z}}, \mathbf{p}^{\mathbf{z}}, \mathbf{q})$ and we obtain an algorithm optimizing $\mathcal{I}(\mathbf{x}^{\mathbf{zw}}, \mathbf{p}^{\mathbf{z}}, \mathbf{q}^{\mathbf{w}})$. The principal steps are described in algorithm 5.2.

**Algorithm 5.2** CROEUC

**input:** x, $g$, $m$
**initialization:** z, w
**repeat**
  $x^{\mathbf{w}}_{i\ell} = \frac{1}{w_{.\ell}} \sum_j w_{j\ell} x_{ij}$, $x^{\mathbf{zw}}_{k\ell} = \frac{1}{z_{.k}} \sum_i z_{ik} x^{\mathbf{w}}_{i\ell}$
  **repeat**
    **step 1.** $z_i = \arg\min_k \sum_\ell w_{.\ell}(x^{\mathbf{w}}_{i\ell} - x^{\mathbf{zw}}_{k\ell})^2$
    **step 2.** $x^{\mathbf{zw}}_{k\ell} = \frac{\sum_i z_{ik} x^{\mathbf{w}}_{i\ell}}{z_{.k}}$
  **until** convergence
  $x^{\mathbf{z}}_{kj} = \frac{1}{z_{.k}} \sum_i z_{ik} x_{ij}$, $x^{\mathbf{zw}}_{k\ell} = \frac{1}{w_{.\ell}} \sum_j z_{j\ell} x^{\mathbf{z}}_{kj}$
  **repeat**
    **step 3.** $w_j = \arg\min_\ell \sum_k z_{.k}(x^{\mathbf{z}}_{kj} - x^{\mathbf{zw}}_{k\ell})^2$
    **step 4.** $x^{\mathbf{zw}}_{k\ell} = \frac{\sum_j w_{j\ell} x^{\mathbf{z}}_{kj}}{w_{.\ell}}$
  **until** convergence
**until** convergence
**return** z, w

REMARK 5.2.– Steps 1 and 2 can be seen as a $k$-means applied on $\mathbf{x}^{\mathbf{w}}$ using the Euclidean distance weighted by $(w_{.1}, \ldots, w_{.m})$

and the mean values of each block. Similarly, steps 3 and 4 can be seen as $k$-means applied on $\mathrm{x}^{\mathbf{z}}$ using the Euclidean distance weighted by $(z_{.1},\ldots,z_{.g})$.

As we have seen in the previous chapters, co-clustering can be embedded in the probabilistic approach. In this chapter, two probabilistic models are studied and we begin by considering the latent block model.

## 5.2. Gaussian latent block model

### 5.2.1. *The model*

Using the latent block model described in section 2.3 of Chapter 2 in the continuous situation, and assuming that for each block $k\ell$ the values $x_{ij}$ are distributed according to a Gaussian distribution

$$\mathcal{N}(\mu_{k\ell},\sigma^2_{k\ell}) \qquad \text{with} \quad \mu_{k\ell}\in\mathbb{R} \quad \text{and} \quad \sigma^2_{k\ell}\in\mathbb{R}^+,$$

we obtain the Gaussian latent block model with the following probability density function (pdf) $f(\mathrm{x};\boldsymbol{\theta})$ taking the form

$$\sum_{(\mathbf{z},\mathbf{w})\in\mathcal{Z}\times\mathcal{W}}\prod_{i,k}\pi_k^{z_{ik}}\prod_{j,\ell}\rho_\ell^{w_{j\ell}}\prod_{i,j,k,\ell} \times\left(\frac{1}{\sqrt{2\pi\sigma^2_{k\ell}}}\exp\left\{\frac{1}{2\sigma^2_{k\ell}}(x_{ij}-\mu_{k\ell})^2\right\}\right)^{z_{ik}w_{j\ell}} \qquad [5.4]$$

parameterized by $\boldsymbol{\theta} = (\boldsymbol{\pi},\boldsymbol{\rho},\boldsymbol{\alpha})$ where $\boldsymbol{\pi} = (\pi_1,\ldots,\pi_g)$, $\boldsymbol{\rho}=(\rho_1,\ldots,\rho_m)$ and $\boldsymbol{\alpha}=(\mu_{11},\ldots,\mu_{gm},\sigma^2_{11},\ldots,\sigma^2_{gm})$. With this model, the complete-data log-likelihood is, up to the constant $-\frac{nd}{2}\log 2\pi$, given by

$$\mathrm{L_C}(\mathbf{z},\mathbf{w},\boldsymbol{\theta}) = \sum_{k,\ell} z_{ik}\log\pi_k + \sum_{j,\ell} w_{j\ell}\log\rho_\ell - \frac{1}{2}\sum_{k,\ell}\left(z_{.k}w_{.\ell}\log\sigma^2_{k\ell} + \frac{1}{\sigma^2_{k\ell}}\sum_{i,j} z_{ik}w_{j\ell}(x_{ij}-\mu_{k\ell})^2\right)$$

and therefore the following fuzzy clustering criterion can be deduced

$$\mathrm{F_C}(\widetilde{\mathbf{z}}, \widetilde{\mathbf{w}}; \boldsymbol{\theta}) = \mathrm{L_C}(\widetilde{\mathbf{z}}, \widetilde{\mathbf{w}}, \boldsymbol{\theta}) + \mathrm{H}(\widetilde{\mathbf{z}}) + \mathrm{H}(\widetilde{\mathbf{w}}). \qquad [5.5]$$

### 5.2.2. *Gaussian* LBVEM *and* LBCEM *algorithms*

Taking into account the definition of the pdf $f(x_{ij}; \mu_{k\ell}, \alpha_{k\ell})$, the variational approximation of the expectation–maximization (EM) algorithm described in section 2.4.1 of Chapter 2 can be completed by the computation of the block parameters $\mu_{k\ell}$ and $\sigma^2_{k\ell}$: for all $k, \ell$, the parameters $\mu_{k\ell}$ and $\sigma^2_{k\ell}$ are obtained by the maximization of

$$-\frac{1}{2}\sum_{k,\ell}\left(\widetilde{z}_{.k}\widetilde{\mathrm{w}}_{.\ell}\log\sigma^2_{k\ell} + \frac{1}{\sigma^2_{k\ell}}\sum_{i,j}\widetilde{z}_{ik}\widetilde{\mathrm{w}}_{j\ell}(x_{ij}-\mu_{k\ell})^2\right).$$

We can easily deduce that the parameter $\mu_{k\ell}$, obtained by the minimization of $\sum_{ij}\widetilde{z}_{ik}\widetilde{\mathrm{w}}_{j\ell}(x_{ij}-\mu_{k\ell})^2$, is

$$\mu_{k\ell} = \frac{\sum_{ij}\widetilde{z}_{ik}\widetilde{\mathrm{w}}_{j\ell}x_{ij}}{\widetilde{z}_{.k}\widetilde{\mathrm{w}}_{.\ell}}$$

and that the parameter $\sigma^2_{k\ell}$, obtained by the minimization of

$$\widetilde{z}_{.k}\widetilde{\mathrm{w}}_{.\ell}\log\sigma^2_{k\ell} + \frac{1}{\sigma^2_{k\ell}}\sum_{ij}\widetilde{z}_{ik}\widetilde{\mathrm{w}}_{j\ell}(x_{ij}-\mu_{k\ell})^2,$$

is

$$\sigma^2_{k\ell} = \frac{\sum_{ij}\widetilde{z}_{ik}\widetilde{\mathrm{w}}_{j\ell}(x_{ij}-\mu_{k\ell})^2}{\widetilde{z}_{.k}\widetilde{\mathrm{w}}_{.\ell}} = \frac{\sum_{ij}\widetilde{z}_{ik}\widetilde{\mathrm{w}}_{j\ell}x^2_{ij}}{\widetilde{z}_{.k}\widetilde{\mathrm{w}}_{.\ell}} - \mu^2_{k\ell}.$$

The principal steps of Gaussian LBVEM are described in algorithm 5.3. In a similar way, the maximization of the $\mathrm{L_C}$ criterion can be performed by the Gaussian LBCEM algorithm presented in algorithm 5.4.

**Algorithm 5.3** Gaussian LBVEM

**input:** x, $g$, $m$

**initialization:** $\widetilde{\mathrm{z}}$, $\widetilde{\mathrm{w}}$, $\pi_k = \frac{\widetilde{z}_{.k}}{n}$, $\rho_\ell = \frac{\widetilde{\mathrm{w}}_{.\ell}}{d}$, $\mu_{k\ell} = \frac{x_{k\ell}^{\widetilde{\mathbf{z}}\widetilde{\mathbf{w}}}}{\widetilde{z}_{.k}\widetilde{\mathrm{w}}_{.\ell}}$. $\sigma_{k\ell}^2 = \frac{\sum_{ij}\widetilde{z}_{ik}\widetilde{\mathrm{w}}_{j\ell}x_{ij}^2}{\widetilde{z}_{.k}\widetilde{\mathrm{w}}_{.\ell}} - \mu_{k\ell}^2$

**repeat**

$x_{i\ell}^{\widetilde{\mathbf{w}}} = \frac{1}{\widetilde{\mathrm{w}}_{.\ell}}\sum_j \widetilde{\mathrm{w}}_{j\ell}x_{ij}$, $u_{i\ell}^{\widetilde{\mathbf{w}}} = \frac{1}{\widetilde{\mathrm{w}}_{.\ell}}\sum_j \widetilde{\mathrm{w}}_{j\ell}x_{ij}^2$

**repeat**

**step 1.** $\widetilde{z}_{ik} \propto \pi_k \exp\left(-\frac{1}{2}\sum_\ell \widetilde{\mathrm{w}}_{.\ell}\left(\log\sigma_{k\ell}^2 + \frac{u_{i\ell}^{\widetilde{\mathbf{w}}} - 2\mu_{k\ell}x_{i\ell}^{\widetilde{\mathbf{w}}} + \mu_{k\ell}^2}{\sigma_{k\ell}^2}\right)\right)$

**step 2.** $\pi_k = \frac{\widetilde{z}_{.k}}{n}$, $\mu_{k\ell} = \frac{\sum_i \widetilde{z}_{ik}x_{i\ell}^{\widetilde{\mathbf{w}}}}{\widetilde{z}_{.k}}$, $\sigma_{k\ell}^2 = \frac{\sum_i \widetilde{z}_{ik}u_{i\ell}^{\widetilde{\mathbf{w}}}}{\widetilde{z}_{.k}} - \mu_{k\ell}^2$

**until** convergence

$x_{kj}^{\widetilde{\mathbf{z}}} = \frac{1}{\widetilde{z}_{.k}}\sum_i \widetilde{z}_{ik}x_{ij}$, $v_{kj}^{\widetilde{\mathbf{z}}} = \frac{1}{\widetilde{z}_{.k}}\sum_i \widetilde{z}_{ik}x_{ij}^2$

**repeat**

**step 3.** $\widetilde{\mathrm{w}}_{j\ell} \propto \rho_\ell \exp\left(-\frac{1}{2}\sum_k \widetilde{z}_{.k}\left(\log\sigma_{k\ell}^2 + \frac{v_{kj}^{\widetilde{\mathbf{z}}} - 2\mu_{k\ell}x_{kj}^{\widetilde{\mathbf{z}}} + \mu_{k\ell}^2}{\sigma_{k\ell}^2}\right)\right)$

**step 4.** $\rho_\ell = \frac{\widetilde{\mathrm{w}}_{.\ell}}{d}$, $\mu_{k\ell} = \frac{\sum_j \widetilde{\mathrm{w}}_{j\ell}x_{kj}^{\widetilde{\mathbf{z}}}}{\widetilde{\mathrm{w}}_{.\ell}}$, $\sigma_{k\ell}^2 = \frac{\sum_j \widetilde{\mathrm{w}}_{j\ell}v_{kj}^{\widetilde{\mathbf{z}}}}{\widetilde{\mathrm{w}}_{.\ell}} - \mu_{k\ell}^2$

**until** convergence

**until** convergence

**return** $\pi$, $\rho$, $\alpha$

### 5.2.3. *Parsimonious Gaussian latent block models*

As in section 1.7.2 of Chapter 1, a parsimonious model can be defined by imposing constraints on the variances: we obtain the $[\sigma]$ model when the variances depend neither on the row cluster nor on the column cluster, the $[\sigma_k]$ model when the variances depend only on the row cluster, and the $[\sigma^j]$ model when the variances depend only on the column cluster.

This parsimonious parameterization allows us to study the relationship between CROEUC algorithm described in section 5.1 and the latent block model. Indeed, in the simplest case, in the $[\sigma]$ model, given identical proportions ($\pi_k = 1/g$ and $\rho_\ell = 1/m$), the complete-data log-likelihood takes the form

$$\mathrm{L_C}(\mathbf{z},\mathbf{w},\boldsymbol{\alpha}) = -\frac{nd}{2}\log\sigma^2 - \frac{1}{2\sigma^2}\sum_{i,j,k,\ell} z_{ik}w_{j\ell}(x_{ij} - \mu_{k\ell})^2 - n\log g - d\log m$$

and it is easy to see that maximizing $\mathrm{L_C}$ is equivalent to minimizing $\mathrm{W}(\mathbf{z},\mathbf{w})$ where

$$\mathrm{W}(\mathbf{z},\mathbf{w}) = \sum_{i,j,k,\ell} z_{ik}w_{j\ell}(x_{ij} - x_{k\ell}^{\mathbf{zw}})^2$$

is the criterion maximized by the CROEUC algorithm.

---

**Algorithm 5.4** Gaussian LBCEM

---

**input:** $\mathbf{x}$, $g$, $m$

**initialization:** $\mathbf{z}$, $\mathbf{w}$, $\pi_k = \frac{z_{.k}}{n}$ $\rho_\ell = \frac{w_{.\ell}}{d}$, $\mu_{k\ell} = \frac{x_{k\ell}^{\mathbf{zw}}}{z_{.k}w_{.\ell}}$. $\sigma_{k\ell}^2 = \frac{\sum_{ij} z_{ik}w_{j\ell}x_{ij}^2}{z_{.k}w_{.\ell}} - \mu_{k\ell}^2$

**repeat**

  $x_{i\ell}^{\mathbf{w}} = \frac{1}{w_{.\ell}}\sum_j w_{j\ell}x_{ij}$, $u_{i\ell}^{\mathbf{w}} = \frac{1}{w_{.\ell}}\sum_j w_{j\ell}x_{ij}^2$

  **repeat**

    **step 1.** $z_i = \arg\max_k \log\pi_k - \frac{1}{2}\sum_\ell w_{.\ell}\left(\log\sigma_{k\ell}^2 + \frac{u_{i\ell}^{\mathbf{w}} - 2\mu_{k\ell}x_{i\ell}^{\mathbf{w}} + \mu_{k\ell}^2}{\sigma_{k\ell}^2}\right)$

    **step 2.** $\pi_k = \frac{z_{.k}}{n}$, $\mu_{k\ell} = \frac{\sum_i z_{ik}x_{i\ell}^{\mathbf{w}}}{z_{.k}}$, $\sigma_{k\ell}^2 = \frac{\sum_i z_{ik}u_{i\ell}^{\mathbf{w}}}{z_{.k}} - \mu_{k\ell}^2$

  **until** convergence

  $x_{kj}^{\mathbf{z}} = \frac{1}{z_{.k}}\sum_i z_{ik}x_{ij}$, $v_{kj}^{\mathbf{z}} = \frac{1}{z_{.k}}\sum_i z_{ik}x_{ij}^2$

  **repeat**

    **step 3.** $w_j = \arg\max_\ell \log\rho_\ell - \frac{1}{2}\sum_k z_{.k}\left(\log\sigma_{k\ell}^2 + \frac{v_{kj}^{\mathbf{z}} - 2\mu_{k\ell}x_{kj}^{\mathbf{z}} + \mu_{k\ell}^2}{\sigma_{k\ell}^2}\right)$

    **step 4.** $\rho_\ell = \frac{w_{.\ell}}{d}$, $\mu_{k\ell} = \frac{\sum_j w_{j\ell}x_{kj}^{\mathbf{z}}}{w_{.\ell}}$, $\sigma_{k\ell}^2 = \frac{\sum_j w_{j\ell}v_{kj}^{\mathbf{z}}}{w_{.\ell}} - \mu_{k\ell}^2$

  **until** convergence

**until** convergence

**return** $\mathbf{z}$, $\mathbf{w}$, $\boldsymbol{\pi}$, $\boldsymbol{\rho}$, $\boldsymbol{\alpha}$

---

We now establish connections between LBCEM and CROEUC. To this end, it suffices to remark that in step 1 of LBCEM, we have

$$z_i = \arg\max_k \log \pi_k - \frac{1}{2}\sum_\ell w_{.\ell}\left(\log \sigma^2_{k\ell} + \frac{u^{\mathbf{w}}_{i\ell} - 2\mu_{k\ell}x^{\mathbf{w}}_{i\ell} + \mu^2_{k\ell}}{\sigma^2_{k\ell}}\right). \quad [5.6]$$

For the $[\sigma]$ model, this leads to

$$z_i = \arg\min_k \sum_\ell w_{.\ell}\left(u^{\mathbf{w}}_{i\ell} - 2\mu_{k\ell}x^{\mathbf{w}}_{i\ell} + \mu^2_{k\ell}\right)$$

or

$$z_i = \arg\min_k \sum_\ell \left(w_{.\ell}(u^{\mathbf{w}}_{i\ell} - (x^{\mathbf{w}}_{i\ell})^2) + w_{.\ell}(x^{\mathbf{w}}_{i\ell} - \mu_{k\ell})^2\right)$$

and finally, since the first term does not depend on $k$, $z_{ik}$ becomes

$$z_i = \arg\min_k \sum_\ell w_{.\ell}(x^{\mathbf{w}}_{i\ell} - \mu_{k\ell})^2.$$

In the same way, we can prove that in step 3 of LBCEM, we have

$$w_j = \arg\min_\ell \sum_k z_{.k}(x^{\mathbf{z}}_{kj} - \mu_{k\ell})^2.$$

Hence, it clearly appears that the CROEUC algorithm is just a particular version of the Gaussian LBCEM algorithm for the parsimonious model $[\sigma^2]$ assuming equal proportions.

## 5.3. Illustrative example

To illustrate the LBVEM algorithm with the $[\sigma_{k\ell}]$ model, we use the Amiard fishes data set presented in [CAI 76, p. 277].

Although the size of this data set is small, it is interesting to note that the number of variables is large compared to the small number of individuals. For the sake of brevity, we will only note that the data consist of a set of 24 fish placed in a radioactive environment. The first nine variables (ey: radioactivity of eyes, gi: radioactivity of gills, ca: radioactivity of cappings, fi: radioactivity of fins, le: radioactivity of lever, gt: radioactivity of the gastrointestinal trac, ki: radioactivity of kidney, sc: radioactivity of scales, mu: radioactivity of muscles) correspond to the level of radioactivity in different organs and the remaining variables (wgt: weight, l: length, sl: standard length, whe: width of head, w: width, wsn: width of snout, dey: diameter of eyes) correspond to different sizes. The 17th fish died during the experiment. Tables 5.2 and 5.3 present the standardized data matrix and the correlation data matrix.

We run the LBVEM algorithm with $(g, m) = (5, 3)$ that seems most relevant, we retain the best co-clustering and we report the clusters in Table 5.4 and on the two planes obtained by PCA and depicted in Figures 5.2 and 5.3. In Table 5.5, we present the mean and variance of each block allowing us to evaluate their degree of homogeneity. We can observe, for instance, the opposition between row clusters 3 and 4 that are characterized by column clusters 1 (level radioactivity) and 2 (size). Row cluster 3 has the smaller mean value for the level of radioactivity in different organs except *rki* and a higher mean for different sizes; while row cluster 4 presents the higher values for level of radioactivity (column cluster 1) and smaller mean for different sizes (column cluster 2). Note that this opposition is confirmed in Figure 5.2. Furthermore, we observe the opposition between row clusters 1 and 2 due to column cluster 1 (size). Row cluster 1 presents the smallest mean for column cluster 1 while row cluster 2 presents the higher mean for the level of radioactivity. Row cluster 5 does not seem relevant, its averages are fairly close to 0 and also located near the origin (Figure 5.2). Finally, column cluster 3 (radioactivity of kidney) seems relevant mainly for row cluster 4.

| Ind. | rey | rgi | rca | rfi | rle | rgt | rki | rsc | rmu | wgt | l | sl | whe | w | wsn | dey |
|---|---|---|---|---|---|---|---|---|---|---|---|---|---|---|---|---|
| 1 | -0.724 | -0.776 | -0.598 | -0.650 | -1.228 | -0.794 | 2.167 | -0.737 | -0.842 | 1.894 | 1.314 | 1.677 | 2.335 | 1.677 | 1.748 | 1.308 |
| 2 | -0.857 | -1.202 | -0.950 | -1.100 | 0.108 | -0.651 | 0.286 | -0.941 | -0.470 | 1.515 | 1.649 | 1.740 | 1.294 | 1.012 | 0.960 | 0.271 |
| 3 | -1.257 | -1.124 | -0.516 | -0.785 | -0.985 | -0.346 | -0.027 | -0.836 | -0.470 | 1.780 | 1.649 | 1.740 | 1.294 | 1.234 | 1.354 | 1.308 |
| 4 | -1.124 | -0.679 | -0.936 | -1.111 | -1.167 | 0.110 | 0.286 | -0.983 | -0.470 | 1.932 | 1.928 | 1.804 | 1.918 | 1.899 | 0.565 | 1.308 |
| 5 | -0.990 | -0.892 | -0.571 | -0.729 | -0.803 | 0.029 | -1.595 | -0.254 | -0.842 | -0.952 | -1.256 | -1.383 | -1.203 | -0.540 | -1.799 | -0.767 |
| 6 | -0.990 | -1.144 | -0.733 | -0.594 | -0.620 | -0.643 | -0.341 | -0.685 | -0.842 | -0.876 | -0.697 | -0.682 | -0.995 | -0.983 | -1.011 | -0.767 |
| 7 | -1.124 | -1.124 | -0.990 | -0.875 | -0.681 | -0.701 | -1.595 | -0.642 | -0.842 | -0.876 | -0.809 | -0.937 | -0.579 | -0.761 | -1.011 | -0.767 |
| 8 | -0.591 | -0.504 | -0.855 | -0.785 | -0.438 | -0.678 | 0.286 | -0.742 | 0.275 | -1.331 | -0.809 | -0.363 | -0.787 | -1.870 | -1.405 | -1.805 |
| 9 | -0.324 | -0.485 | -0.611 | -0.111 | 0.898 | -0.346 | -0.027 | -0.609 | -0.097 | -0.383 | -0.474 | -0.427 | -0.579 | -0.096 | -0.617 | 0.271 |
| 10 | 0.741 | 0.871 | 0.079 | -0.212 | 1.323 | -0.203 | -0.027 | -0.306 | 0.648 | -0.117 | 0.532 | 0.529 | 0.461 | -0.318 | -0.617 | -0.767 |
| 11 | -0.458 | -0.272 | -0.340 | -0.302 | -0.317 | 1.190 | -0.027 | -0.368 | -0.470 | -0.079 | -0.306 | -0.491 | 0.045 | 0.347 | -0.617 | 1.308 |
| 12 | -0.191 | 0.290 | -0.449 | -0.448 | -0.378 | 0.106 | -0.027 | 1.511 | 0.648 | -0.383 | -0.865 | -0.809 | -0.579 | -0.096 | -0.223 | 0.271 |
| 13 | -0.191 | 0.716 | -0.313 | -0.336 | 0.412 | 0.932 | -0.027 | -0.410 | 2.511 | -0.269 | -0.083 | -0.108 | -0.163 | -0.096 | 1.748 | 0.271 |
| 14 | 1.008 | -0.253 | -0.395 | -0.369 | 1.323 | 0.685 | -0.027 | -0.477 | -0.470 | -1.142 | -1.479 | -1.510 | -1.412 | -0.983 | -0.617 | -0.767 |
| 15 | -0.324 | -0.388 | -0.611 | -0.459 | -0.438 | 0.140 | -0.027 | -0.505 | 0.275 | 0.148 | 0.253 | 0.274 | -0.371 | -0.096 | 0.960 | 0.271 |
| 16 | -0.191 | 0.019 | -0.571 | -0.617 | 0.230 | -0.643 | -0.027 | -0.235 | 1.021 | 0.186 | 1.091 | -0.044 | 0.670 | 0.125 | 1.354 | 0.271 |
| 18 | 2.207 | 2.304 | 2.043 | 1.676 | 0.533 | -0.674 | 1.227 | 0.773 | -0.097 | -0.383 | -0.529 | -0.427 | -0.371 | -0.761 | -0.223 | -0.767 |
| 19 | 0.875 | 1.103 | 1.474 | 1.721 | -0.135 | 2.325 | -1.281 | 1.383 | -0.470 | -0.724 | -0.865 | -0.682 | -0.995 | -0.983 | -0.617 | -0.767 |
| 20 | 2.074 | 1.743 | 1.339 | 2.080 | 2.780 | -0.666 | -1.281 | 2.419 | 2.884 | -1.256 | -1.144 | -1.064 | -0.787 | -1.427 | -0.617 | -1.805 |
| 21 | -0.058 | 0.425 | 0.134 | 0.361 | -0.256 | -0.705 | -0.654 | -0.666 | -0.470 | 0.945 | 0.755 | 0.912 | 0.878 | 1.234 | 0.565 | 1.308 |
| 22 | 0.875 | 1.065 | 1.989 | 1.316 | -0.924 | -0.693 | 0.600 | 1.856 | -0.097 | 0.035 | -0.027 | 0.338 | -0.163 | 1.012 | 0.171 | -0.767 |
| 23 | 1.141 | 1.103 | 1.651 | 1.608 | 1.444 | 2.892 | 2.481 | 1.232 | -0.470 | -0.003 | 0.197 | -0.172 | -0.163 | -0.096 | 0.171 | 0.271 |
| 24 | 0.475 | -0.795 | 0.730 | 0.721 | -0.681 | -0.666 | -0.341 | 0.224 | -0.842 | 0.338 | -0.027 | 0.083 | 0.253 | 0.569 | -0.223 | 1.308 |

**Table 5.2.** *Standardized Amiard fishes data set*

| | rey | rgi | rca | rfi | rle | rgt | rki | rsc | rmu | wgt | l | sl | whe | w | wsn | dey |
|---|---|---|---|---|---|---|---|---|---|---|---|---|---|---|---|---|
| rey | 1.0 | 0.9 | 0.9 | 0.9 | 0.7 | 0.2 | 0.2 | 0.7 | 0.4 | −0.4 | −0.4 | −0.4 | −0.3 | −0.4 | −0.1 | −0.4 |
| rgi | 0.9 | 1.0 | 0.8 | 0.8 | 0.6 | 0.3 | 0.2 | 0.7 | 0.5 | −0.3 | −0.3 | −0.3 | −0.2 | −0.3 | −0.0 | −0.4 |
| rca | 0.9 | 0.8 | 1.0 | 1.0 | 0.4 | 0.3 | 0.2 | 0.8 | 0.1 | −0.2 | −0.2 | −0.2 | −0.2 | −0.1 | −0.1 | −0.3 |
| rfi | 0.9 | 0.8 | 1.0 | 1.0 | 0.5 | 0.3 | 0.1 | 0.8 | 0.2 | −0.3 | −0.3 | −0.3 | −0.3 | −0.3 | −0.2 | −0.3 |
| rle | 0.7 | 0.6 | 0.4 | 0.5 | 1.0 | 0.2 | 0.0 | 0.4 | 0.6 | −0.4 | −0.3 | −0.4 | −0.4 | −0.5 | −0.2 | −0.4 |
| rgt | 0.2 | 0.3 | 0.3 | 0.3 | 0.2 | 1.0 | 0.2 | 0.3 | −0.0 | −0.2 | −0.2 | −0.2 | −0.3 | −0.1 | −0.0 | 0.1 |
| rki | 0.2 | 0.2 | 0.2 | 0.1 | 0.0 | 0.2 | 1.0 | −0.0 | −0.1 | 0.4 | 0.4 | 0.4 | 0.4 | 0.4 | 0.5 | 0.3 |
| rsc | 0.7 | 0.7 | 0.8 | 0.8 | 0.4 | 0.3 | −0.0 | 1.0 | 0.4 | −0.4 | −0.4 | −0.4 | −0.4 | −0.3 | −0.2 | −0.4 |
| rmu | 0.4 | 0.5 | 0.1 | 0.2 | 0.6 | −0.0 | −0.1 | 0.4 | 1.0 | −0.3 | −0.1 | −0.2 | −0.2 | −0.3 | 0.2 | −0.3 |
| wgt | −0.4 | −0.3 | −0.2 | −0.3 | −0.4 | −0.2 | 0.4 | −0.4 | −0.3 | 1.0 | 0.9 | 0.9 | 0.9 | 0.9 | 0.7 | 0.8 |
| l | −0.4 | −0.3 | −0.2 | −0.3 | −0.3 | −0.2 | 0.4 | −0.4 | −0.1 | 0.9 | 1.0 | 1.0 | 0.9 | 0.8 | 0.8 | 0.7 |
| sl | −0.4 | −0.3 | −0.2 | −0.3 | −0.4 | −0.2 | 0.4 | −0.4 | −0.2 | 0.9 | 1.0 | 1.0 | 0.9 | 0.8 | 0.7 | 0.6 |
| whe | −0.3 | −0.2 | −0.2 | −0.3 | −0.4 | −0.3 | 0.4 | −0.4 | −0.2 | 0.9 | 0.9 | 0.9 | 1.0 | 0.9 | 0.7 | 0.7 |
| w | −0.4 | −0.3 | −0.1 | −0.3 | −0.5 | −0.1 | 0.4 | −0.3 | −0.3 | 0.9 | 0.8 | 0.8 | 0.9 | 1.0 | 0.7 | 0.8 |
| wsn | −0.1 | −0.0 | −0.1 | −0.2 | −0.2 | −0.0 | 0.5 | −0.2 | 0.2 | 0.7 | 0.8 | 0.7 | 0.7 | 0.7 | 1.0 | 0.6 |
| dey | −0.4 | −0.4 | −0.3 | −0.3 | −0.4 | 0.1 | 0.3 | −0.4 | −0.3 | 0.8 | 0.7 | 0.6 | 0.7 | 0.8 | 0.6 | 1.0 |

**Table 5.3.** *Correlation matrix*

| Row clusters | Column clusters |
|---|---|
| cluster 1 (4 obs.) : 5 6 7 8 | cluster 1 (7 obs.) : rey rgi rca rfi rle rgt rsc rmu |
| cluster 2 (4 obs.) : 18 22 23 24 | cluster 2 (8 obs.) : wgt l sl whe w wsn dey |
| cluster 3 (5 obs.) : 1 2 3 4 21 | cluster 3 (1 obs.) : rki |
| cluster 4 (2 obs.) : 19 20 | |
| cluster 5 (8 obs.) : 9 10 11 12 13 14 15 16 | |

**Table 5.4.** *Row and column clusters*

| $\mu_{k\ell}$ | 1 | 2 | 3 |
|---|---|---|---|
| 1 | -0.71 | -1.00 | -0.81 |
| 2 | 1.43 | -0.98 | -1.28 |
| 3 | -0.66 | 1.39 | 0.41 |
| 4 | 0.75 | -0.01 | 0.99 |
| 5 | 0.03 | -0.12 | -0.034 |

| $\sigma^2_{k\ell}$ | 1 | 2 | 3 |
|---|---|---|---|
| 1 | 0.09 | 0.14 | 0.66 |
| 2 | 1.10 | 0.11 | 0.00 |
| 3 | 0.19 | 0.21 | 0.89 |
| 4 | 1.14 | 0.22 | 1.05 |
| 5 | 0.43 | 0.44 | 0.00 |

**Table 5.5.** *Means and variances of each block*

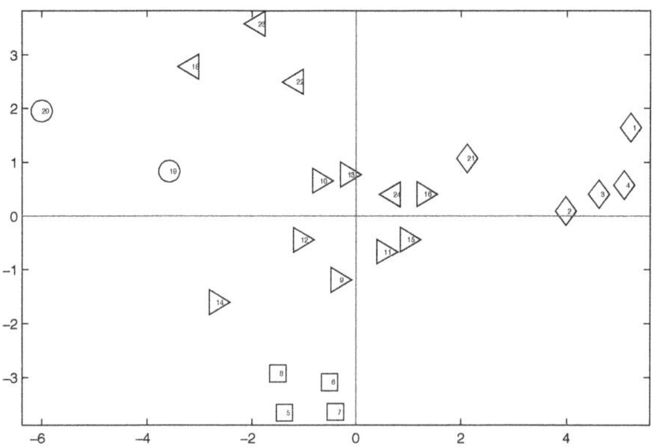


**Figure 5.2.** *Projection of the columns into the factorial plane spawned by the first and second axes that account for 71.07% of variance*

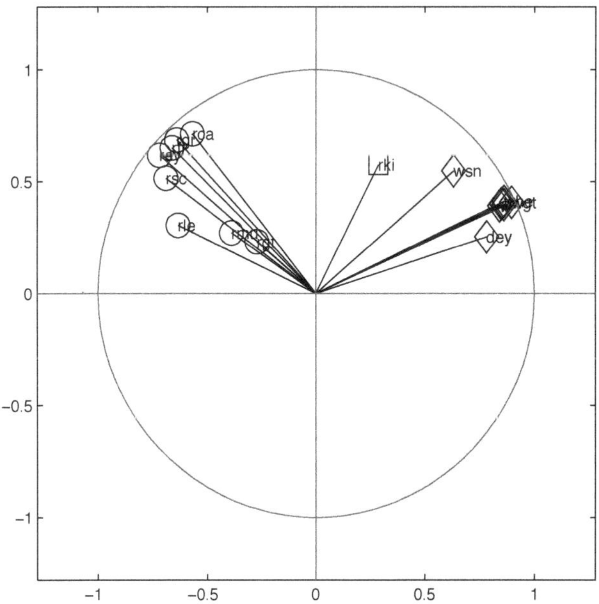


**Figure 5.3.** *Projection of the columns into the factorial plane spawned by the first and second axes that account for 71.07% of variance*

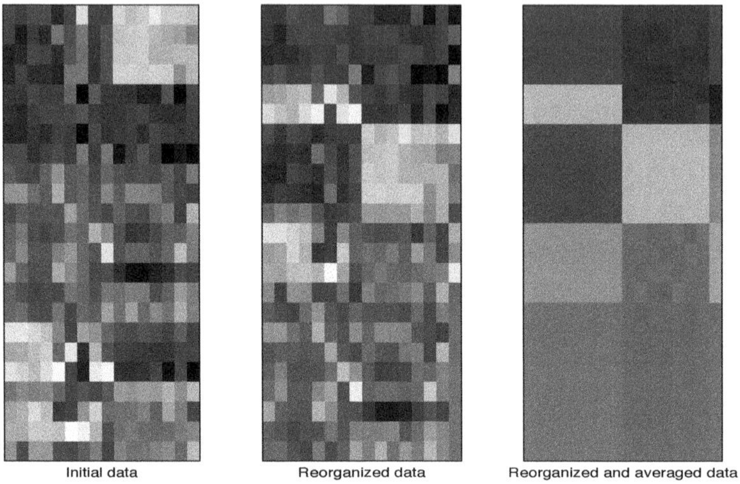


**Figure 5.4.** *Visualization of initial data and results after co-clustering*

## 5.4. Gaussian block mixture model

The Gaussian latent block model, being symmetric in both rows and columns, can appear less adapted to continuous data. In this case, the data matrix presents a dissymmetry: rows and columns that correspond to entities of different types will not be handled in the same way, contrary to what we have done in the two previous chapters for binary and contingency tables. Let us recall that we encounter the same problem with PCA where the individuals and the variables are not treated in a symmetrical way, which is not the case of *correspondence analysis* that treats the rows and the columns of a contingency table in the same way.

To overcome this difficulty, another model based on the classical Gaussian mixture model can be proposed. In this section, we describe this model and establish some connections with the latent block model.

### 5.4.1. *The model*

Hereafter, we use a classical mixture model in which the partition w of the variables is considered as a parameter of the model [NAD 10]. The pdf is therefore

$$f(\mathbf{x}_i;\boldsymbol{\theta}) = \sum_k \pi_k f(\mathbf{x}_i;\mathbf{w},\boldsymbol{\alpha}),$$

with $f(\mathbf{x}_i;\mathbf{w},\boldsymbol{\alpha}) = \prod_{j,\ell}\left(\frac{1}{\sqrt{2\pi\sigma^2_{k\ell}}} e^{-\frac{1}{2\sigma^2_{k\ell}}(x_{ij}-a_{k\ell})^2}\right)^{w_{j\ell}}$. The unknown parameter $\boldsymbol{\theta}$ is now formed by $\pi$, w and $\alpha$ where $\alpha = (\mathrm{a}, \Sigma)$ with a and $\Sigma$ being $g \times m$ matrices representing the means and the variances of blocks

$$\mathbf{a} = \begin{pmatrix} a_{11} & \dots & a_{1m} \\ \vdots & \ddots & \vdots \\ a_{g1} & \dots & a_{gm} \end{pmatrix}, \Sigma = \begin{pmatrix} \sigma^2_{11} & \dots & \sigma^2_{1m} \\ \vdots & \ddots & \vdots \\ \sigma^2_{g1} & \dots & \sigma^2_{gm} \end{pmatrix},$$

or

$$\boldsymbol{\alpha} = \begin{pmatrix} (a_{11},\sigma^2_{11}) & \dots & (a_{1m},\sigma^2_{1m}) \\ \vdots & \ddots & \vdots \\ (a_{g1},\sigma^2_{g1}) & \dots & (a_{gm},\sigma^2_{gm}) \end{pmatrix}.$$

This model can be viewed as a Gaussian mixture model with constraints on the $g$ mean vectors and $g$ variance matrices. For each component $k$, the $(d \times 1)$ mean vector $\mathrm{a}_k$ takes this form

$$(a_{k1},\dots,a_{k1},a_{k2},\dots,a_{k2},\dots,a_{km},\dots,a_{km})^T,$$

where each $a_{k\ell}$ is repeated $w_\ell$ times. In the same manner, the variance matrix $\Sigma_k$ is a diagonal $(d \times d)$ matrix defined by

$$Diag(\sigma^2_{k1},\dots,\sigma^2_{k1},\sigma^2_{k2},\dots,\sigma^2_{k2},\dots,\sigma^2_{km},\dots,\sigma^2_{km}),$$

where each variance $\sigma_{k\ell}^2$ is repeated $w_\ell$ times. When for each component $k$ the variances are assumed to be equal to $\sigma_k^2$, $\Sigma_k$ becomes $\sigma_k^2 I$. This is a parsimonious model, as opposed to a spherical Gaussian mixture model. The number of parameters is equal to $g + 2(g * m)$ instead of $g + 2(g * d)$. Hence, it is more adapted when $n$ is much smaller than $d$, a classical situation in bioinformatics.

### 5.4.2. GBEM *algorithm*

Setting this model under the maximum likelihood approach, the EM algorithm can be used to estimate the parameters. The complete-data log-likelihood

$$f(x;\boldsymbol{\theta},\mathbf{z}) = \sum_{i,k} z_{ik} \log\left(\pi_k f(\mathbf{x}_i;\mathbf{w},\boldsymbol{\alpha})\right)$$

takes, up to the constant $-\frac{nd}{2}\log 2\pi$, the following form

$$\mathrm{L_C}(\mathbf{z};\boldsymbol{\theta}) = \sum_k z_{.k}\log\pi_k - \frac{1}{2}\sum_{i,j,k,\ell} z_{ik}w_{j\ell}\left(\log\sigma_{k\ell}^2 + \frac{(x_{ij}-a_{k\ell})^2}{\sigma_{k\ell}^2}\right).$$

Using the interpretation of Hathaway [HAT 86] described in section 1.4.5 of Chapter 1, the EM algorithm can be seen as the alternating optimization algorithm of the fuzzy criterion

$$\mathrm{F_C}(\widetilde{\mathbf{z}},\boldsymbol{\theta}) = \mathrm{L_C}(\boldsymbol{\theta},\widetilde{\mathbf{z}}) + \mathrm{H}(\widetilde{\mathbf{z}}),$$

where $\mathrm{H}(\widetilde{\mathbf{z}}) = -\sum_{i,k}\widetilde{z}_{ik}\log\widetilde{z}_{ik}$ is the entropy of the distribution $\widetilde{\mathbf{z}}$. Therefore, the EM algorithm, denoted by Gaussian block EM algorithm (GBEM), alternates the following steps.

– *E step:* the maximization for fixed $\theta$ reduces to the computation of the conditional probabilities. Each probability

$\widetilde{z}_{ik}$ is proportional to $\pi_k f(\mathbf{x}_i; \mathbf{w}, \boldsymbol{\alpha})$ whose logarithm takes the form

$$\log \pi_k - \frac{1}{2}\sum_{\ell}\left(w_\ell \log \sigma^2_{k\ell} + \frac{(e^{\mathbf{w}}_{i\ell} + w_\ell(x^{\mathbf{w}}_{i\ell} - a_{k\ell})^2)}{\sigma^2_{k\ell}}\right),$$

with $x^{\mathbf{w}}_{i\ell} = \frac{\sum_j w_{j\ell}x_{ij}}{w_\ell}$ and $e^{\mathbf{w}}_{i\ell} = \sum_j w_{j\ell}(x_{ij} - x^{\mathbf{w}}_{i\ell})^2$.

– *M step:* the maximization in $\boldsymbol{\theta}$ of $\mathrm{L_C}(\boldsymbol{\theta}, \widetilde{\mathbf{z}})$ is not straightforward. We can use the generalized EM algorithm (GEM) for which the M step requires $\boldsymbol{\theta}$ to be chosen such that $\mathrm{L_C}$ is increased rather than maximized over all $\boldsymbol{\theta}$. The maximization of $\sum_k \widetilde{z}_{.k}\log\pi_k$ leads to $\pi_k = \frac{\widetilde{z}_{.k}}{n}$ and to decrease

$$h(\mathbf{w}, \boldsymbol{\alpha}) = \sum_{i,j,k,\ell} \widetilde{z}_{ik}w_{j\ell}\left(\log \sigma^2_{k\ell} + \frac{(x_{ij} - a_{k\ell})^2}{\sigma^2_{k\ell}}\right)$$

the following alternated minimizations can be used:

1) Computation of w given $\alpha$: this step consists of minimizing $h$ with respect to w. As the expression of $h(\mathbf{w}, \boldsymbol{\alpha})$ can be written as $\sum_{j,\ell} w_{j\ell}t_{j\ell}$ where

$$t_{j\ell} = \sum_k\left(\widetilde{z}_{.k}\log\sigma^2_{k\ell} + \frac{1}{\sigma^2_{k\ell}}\left(f^{\widetilde{\mathbf{z}}}_{kj} + \widetilde{z}_{.k}(x^{\widetilde{\mathbf{z}}}_{kj} - a_{k\ell})^2\right)\right),$$

with $x^{\widetilde{\mathbf{z}}}_{kj} = \frac{\sum_i \widetilde{z}_{ik}x_{ij}}{\widetilde{z}_{.k}}$ and $f^{\widetilde{\mathbf{z}}}_{kj} = \sum_i \widetilde{z}_{ik}(x_{ij} - x^{\widetilde{\mathbf{z}}}_{kj})^2$, we obtain
$w_j = \arg\min_\ell t_{j\ell}$.

2) Computation of $\alpha$ given w: this step consists of minimizing $h$ with respect to $\alpha$ given w. For all $k$ and $\ell$, the expression to minimize is

$$\widetilde{z}_{.k}w_\ell\log\sigma^2_{k\ell} + \sum_{i,j}\widetilde{z}_{ik}w_{j\ell}\frac{(x_{ij} - a_{k\ell})^2}{\sigma^2_{k\ell}},$$

which leads to

$$a_{k\ell} = \frac{\sum_{ij}\widetilde{z}_{ik}w_{j\ell}x_{ij}}{\widetilde{z}_{.k}w_{.\ell}} \quad \text{and} \quad \sigma^2_{k\ell} = \frac{\sum_{i,j}\widetilde{z}_{ik}w_{j\ell}\left(x_{ij} - a_{k\ell}\right)^2}{\widetilde{z}_{.k}w_\ell}.$$

Note that in the M step, using the terms $x^{\widetilde{\mathbf{z}}}_{kj}$ and $f^{\widetilde{\mathbf{z}}}_{kj}$, it is easy to show that the mean and the variance of each block take, respectively, the following forms

$$a_{k\ell} = \frac{\sum_j w_{j\ell} x_{kj}^{\widetilde{\mathbf{z}}}}{w_\ell} \quad \text{and} \quad \sigma_{k\ell}^2 = \frac{\sum_j w_{j\ell}\left(f_{kj}^{\widetilde{\mathbf{z}}} + \widetilde{z}_{.k}(x_{kj}^{\widetilde{\mathbf{z}}} - a_{k\ell})^2\right)}{\widetilde{z}_{.k} w_\ell}, \qquad [5.7]$$

and computational shortcuts can be performed on a reduced matrix using sufficient statistics $x_{kj}^{\widetilde{\mathbf{z}}}$ and $f_{kj}^{\widetilde{\mathbf{z}}}$. The different steps of GBEM are summarized in algorithm 5.5.

**Algorithm 5.5** GBEM

**input:** x, $g$, $m$

**initialization:** z, w, $\pi_k = \frac{z_{.k}}{n}$, $a_{k\ell} = \frac{\sum_{ij} z_{ik} w_{j\ell} x_{ij}}{z_{.k} w_{.\ell}}$, $\sigma_{k\ell}^2 = \frac{\sum_{i,j} z_{ik} w_{j\ell}(x_{ij} - a_{k\ell})^2}{z_{.k} w_\ell}$

**repeat**

  **E-step:** $x_{i\ell}^{\mathbf{w}} = \frac{\sum_j w_{j\ell} x_{ij}}{w_\ell}$, $e_{i\ell}^{\mathbf{w}} = \sum_j w_{j\ell}(x_{ij} - x_{i\ell}^{\mathbf{w}})^2$,

  $\widetilde{z}_{ik} \propto \log \pi_k - \frac{1}{2}\sum_\ell \left(w_\ell \log \sigma_{k\ell}^2 + \frac{(e_{i\ell}^{\mathbf{w}} + w_\ell(x_{i\ell}^{\mathbf{w}} - a_{k\ell})^2)}{\sigma_{k\ell}^2}\right)$

  **M-step:** $\pi_k = \frac{\widetilde{z}_{.k}}{n}$, $x_{kj}^{\widetilde{\mathbf{z}}} = \frac{\sum_i \widetilde{z}_{ik} x_{ij}}{\widetilde{z}_{.k}}$, $f_{kj}^{\widetilde{\mathbf{z}}} = \sum_i \widetilde{z}_{ik}(x_{ij} - x_{kj}^{\widetilde{\mathbf{z}}})^2$

  **repeat**

    **Step a:** $w_j = \arg\min_\ell \sum_k \left(\widetilde{z}_{.k} \log \sigma_{k\ell}^2 + \frac{1}{\sigma_{k\ell}^2}\left(f_{kj}^{\widetilde{\mathbf{z}}} + \widetilde{z}_{.k}(x_{kj}^{\widetilde{\mathbf{z}}} - a_{k\ell})^2\right)\right)$

    **Step b:** $a_{k\ell} = \frac{\sum_j w_{j\ell} x_{kj}^{\widetilde{\mathbf{z}}}}{w_\ell}$, $\sigma_{k\ell}^2 = \frac{\sum_j w_{j\ell}\left(f_{kj}^{\widetilde{\mathbf{z}}} + \widetilde{z}_{.k}(x_{kj}^{\widetilde{\mathbf{z}}} - a_{k\ell})^2\right)}{\widetilde{z}_{.k} w_\ell}$

  **until** convergence

**until** convergence

**return** $\pi$, $\alpha$

Regarding the context of clustering with the maximum likelihood approach, after we estimate parameter $\boldsymbol{\theta}$, we can give a probabilistic clustering of the $n$ individuals in terms of their fitted posterior probabilities of component membership $\widetilde{z}_{ik}$ obtained at the convergence of EM. Therefore, we can obtain a partition by using a classification step that assigns each individual to the component of the mixture for which it has the highest posterior probability of belonging. With the optimal w partition, we therefore obtain a co-clustering where a partition of individuals is characterized by a partition of variables. The GBEM algorithm can be viewed as a soft algorithm to cluster the set of individuals and the set of variables simultaneously.

A hard version called classification GBEM can be performed by replacing the maximization of $\mathrm{L}(\boldsymbol{\theta})$ by the maximization of $\mathrm{L_C}(\mathbf{z}, \mathbf{w}; \boldsymbol{\theta})$. The main modifications concern the conditional maximization of complete-data log-likelihoods with respect to w given z and $\boldsymbol{\theta}$, and with respect to $\boldsymbol{\theta}$ given z and w. This leads us to convert the posterior probabilities $z_{ik}$s into a discrete classification

$$z_{ik} = 1 \quad \text{if} \quad k = \arg\max_{k'=1,\ldots,g} \tilde{z}_{ik'} \quad \text{and} \quad z_{ik} = 0 \quad \text{otherwise}$$

in a C step before performing the M step based this time on the clusters. It can also be remarked that this algorithm is the Gaussian LBCEM algorithm precisely when the column proportions $\rho_\ell$ are equal to $1/m$.

## 5.5. Numerical experiments

In these first experiments, we consider the model where all blocks have the same variance and the proportions of clusters are equal. We have chosen this restriction in order to evaluate the different algorithms in the same condition. First, to demonstrate the advantage of GBEM, we compared its performances with classical EM on the diagonal Gaussian model ignoring the clustering of variables. Second, we evaluated GBEM when the number of columns was higher the number of rows.

### 5.5.1. GBEM *versus* CROEUC *and EM*

To illustrate the behavior of GBEM, we selected $1{,}000 \times 50$ data arising from $3 \times 2$ component mixture models corresponding to three degrees of overlap of the clusters: well separated, moderately separated and poorly separated. The concept of cluster separation for our model is difficult to visualize, but the degree of overlap can be measured by the true error rate, approximated by comparing the partitions simulated with those that we obtained by applying a classification step. From our numerical experiments, we present only three situations corresponding to three levels of overlap degrees: M1 for well-separated clusters ($8.6\%$), M2 for moderately separated clusters ($16\%$) and M3 for poorly separated clusters ($24.8\%$). To compare two partitions $\mathbf{z}$ and $\mathbf{z}'$ with the same number of clusters, the error rate or the proportions of misclassified individuals is denoted by $\delta(\mathbf{z}, \mathbf{z}')$. It can be defined as follows: if $C$ is the confusion matrix between the two partitions, relabel the components of the partition $\mathbf{z}'$ such that the trace of matrix $C$ is maximal (to obtain this maximum value in our experiments, we enumerate all possible relabellings), then compute

$$\delta(\mathbf{z}, \mathbf{z}') = 1 - \frac{1}{n} \sum_{i,k} z_{ik} z'_{ik}.$$

In Table 5.6, we compared the performances of GBEM, EM and CROEUC by using $\delta(\mathbf{z}, \mathbf{z}')$ (in percent) and their execution times recorded from the same initial positions. It is clear that GBEM outperforms EM and CROEUC. On the other hand, GBEM is faster than EM, the rate timeEM/timeGBEM denoted by $t_{EM}/t_{\text{GBEM}}$ is higher than 2. Different Monte Carlo simulations were performed confirming these remarks and also the superiority of GBEM as compared to EM and CROEUC.

| Error (%) | Situation | GBEM | EM | CROEUC | $\frac{t_{EM}}{t_{\text{GBEM}}}$ |
|---|---|---|---|---|---|
| | M1 | 8.5 | 8.6 | 8.5 | 2.01 |
| $\delta(\mathbf{z},\mathbf{z}')$ | M2 | 16.0 | 21.6 | 18.1 | 2.66 |
| | M3 | 19.6 | 35.6 | 24 | 2.16 |

**Table 5.6.** *Comparison between* GBEM, EM *and* CROEUC *($n \times d = 1{,}000 \times 50$)*

### 5.5.2. *Effect of the size of data*

Now, we illustrate the interest of GBEM when $n$ is less than $d$, which is a crucial problem in bioinformatics. As the latent block model is parsimonious, it does not suffer from this problem and therefore offers a good alternative for clustering individuals. Table 5.7 presents the degree of overlap and the error rates $\delta(\mathbf{z},\mathbf{z}')$ for different sizes of $n$. We note that when $n$ is less than $d$, GBEM is always the best even though $n$ approaches $d$ and remains the best when $n$ is greater than $d$.

| $n$ | 20 | 30 | 40 | 400 |
|---|---|---|---|---|
| degree of overlap (%) | 5 | 13 | 15 | 14 |
| $\delta(\mathbf{z},\mathbf{z}')$ for GBEM | 5 | 13 | 17 | 14 |
| $\delta(\mathbf{z},\mathbf{z}')$ for EM | 35 | 26 | 30 | 19 |

**Table 5.7.** GBEM *versus EM when $n < d = 400$*

## 5.6. Conclusion

For continuous data, we have considered their co-clustering under metric and probabilistic approaches. In the latter approach, we have considered the Gaussian latent block model and the Gaussian block mixture model in which the partition of the set of variables is considered as a parameter of the model. We have presented different algorithms and showed their connections. In addition, this probabilistic approach, as for binary data and contingency tables, allows us to give a probabilistic interpretation of an objective function commonly used in co-clustering. In practice, it has been demonstrated with a simple example that co-clustering is very complementary to other exploratory methods such as PCA. To combine co-clustering and visualization, Priam *et al.* [PRI 13] proposed a Gaussian topographic co-clustering model based on the latent block model. Furthermore, when the data matrix is positive, many alternative methods based on non-negative matrix factorization can be proposed [LON 05, DIN 06, YOO 10, LAB 11a] and are frequently evaluated in the document clustering field on contingency tables after converting original data into continuous data using different types of normalization such as tf-idf. Note that this normalization step is crucial for the quality of results and deserves more investigation.

# Bibliography


[AGR 90] AGRESTI A., *Categorical Data Analysis*, Wiley, New York, 1990.

[AGR 98] AGRAWAL R., GEHRKE J., GUNOPULOS D., *et al.*, "Automatic subspace clustering of high dimensional data for data mining applications", *Proceedings of the 1998 ACM SIGMOD International Conference on Management of Data (SIGMOD '98)*, New York, NY, ACM, pp. 94–105, 1998.

[AHM 07] AHMAD W., KHOKHAR A., "cHawk: an efficient biclustering algorithm based on bipartite graph crossing minimization", *International Conference on Very Large Data Bases*, Vienna, Austria, 2007.

[AIR 08] AIROLDI E.M., BLEI D.M., FIENBERG S.E., *et al.*, "Mixed membership stochastic blockmodels", *Journal of Machine Learning Research*, vol. 9, pp. 1823–1856, 2008.

[AIT 76] AITCHISON J., AITKEN C.G.G., "Multivariate binary discrimination by the kernel method", *Biometrika*, vol. 63, pp. 413–420, 1976.

[ALL 09] ALLMAN E., MATIAS C., RHODES J., "Identifiability of parameters in latent structure models with many observed variables", *The Annals of Statistics*, vol. 37, no. 6A, pp. 3099–3132, 2009.

[AMB 02] AMBROISE C., GOVAERT G., "A mixture model approach to datacube clustering (invited)", *26th Annual GFKL (Gesellschaft für Klassifikation)*, University of Mannheim, Germany, 22–24 July 2002.

[AND 73] ANDERBERG M.R., *Cluster Analysis for Applications*, Academic Press, New York, 1973.

[ARA 90] ARABIE P., HUBERT L.J., "The bond energy algorithm revisited", *IEEE Transactions on Systems, Man, and Cybernetics*, vol. 20, pp. 268–274, 1990.

[BAI 97] BAIER D., GAUL W., SCHADER M., "Two-mode overlapping clustering with applications to simultaneous benefit segmentation and market structuring", in KLAR R., OPITZ O. (eds.), *Classification and Knowledge Organization*, Springer, Heidelberg, 1997.

[BAL 67] BALL G.H., HALL D.J., "A clustering technique for summarizing multivariate data", *Behavioral Science*, vol. 12, no. 2, pp. 153–155, 1967.

[BAN 93] BANFIELD J.D., RAFTERY A.E., "Model-based Gaussian and non-Gaussian clustering", *Biometrics*, vol. 49, pp. 803–821, 1993.

[BAN 07] BANERJEE A., DHILLON I., GHOSH J., *et al.*, "A generalized maximum entropy approach to bregman co-clustering and matrix approximation", *Journal of Machine Learning Research*, vol. 8, pp. 1919–1986, 2007.

[BEN 73a] BENZECRI J.P., *L'analyse des données tome 1: la taxinomie*, Dunod, Paris, 1973.

[BEN 73b] BENZECRI J.P., *L'analyse des données tome 2: l'analyse des correspondances*, Dunod, Paris, 1973.

[BEN 03] BEN-DOR A., CHOR B., KARP R., *et al.*, "Discovering local structure in gene expression data: the order-preserving submatrix problem", *Journal of Computational Biology*, vol. 10, no. 3–4, pp. 373–384, 2003.

[BER 80] BERTIN J., "Traitements graphiques et mathématiques. différence fondamentale et complémentarité", *Mathématiques et sciences humaines*, vol. 72, pp. 60–71, 1980.

[BER 03] BERGMANN S., IHMELS J., BARKAI N., "Iterative signature algorithm for the analysis of large-scale gene expression data", *Physics Review E*, vol. 67, pp. 031902-1–031902-18, 2003.

[BEZ 81] BEZDEK J., *Pattern Recognition with Fuzzy Objective Function Algorithms*, Plenum Press, New York, 1981.

[BIE 00] BIERNACKI C., CELEUX G., GOVAERT G., "Assessing a mixture model for clustering with the integrated completed likelihood", *IEEE Transactions on Pattern Analysis and Machine Intelligence*, vol. 22, no. 7, pp. 719–725, 2000.

[BIE 03] BIERNACKI C., CELEUX G., GOVAERT G., "Choosing starting values for the EM algorithm for getting the highest likelihood in multivariate Gaussian mixture models", *Computational Statistics and Data Analysis*, vol. 41, pp. 561–575, 2003.

[BIE 10] BIERNACKI C., CELEUX G., GOVAERT G., "Exact and Monte Carlo calculations of integrated likelihoods for the latent class model", *Journal of Statistical Planning and Inference*, vol. 140, no. 11, pp. 2991–3002, 2010.

[BIS 98] BISHOP C.M., SVENSEN M., WILLIAMS C.K.I., "GTM: the generative topographic mapping", *Neural Computation*, vol. 10, no. 1, pp. 215–234, 1998.

[BIS 12] BISSON G., GRIMAL C., "Co-clustering of multi-view datasets: a parallelizable approach", *IEEE 12th International Conference on Data Mining*, pp. 828–833, 2012.

[BLE 03] BLEI D., NG A., JORDAN M., *et al.*, "Latent dirichlet allocation", *Journal of Machine Learning Research*, vol. 3, pp. 993–1022, 2003.

[BOC 79] BOCK H., "Simultaneous clustering of objects and variables", in TOMASSONE R. (ed.), *Analyse des Données et Informatique*, INRIA, Le Chesnay, France, pp. 187–203, 1979.

[BOC 86] BOCK H., "Loglinear models and entropy clustering methods for qualitative data", in GAULL W., SCHADER M. (eds.), *Classification as a Tool of Research*, North-Holland Publishing Company, Amsterdam, pp. 19–26, 1986.

[BOC 89] BOCK H., "Probabilistic aspects in cluster analysis", in OPITZ O. (ed.), *Conceptual and Numerical Analysis of Data*, Springer-Verlag, Berlin, pp. 12–44, 1989.

[BOC 07] BOCK H., "Clustering methods: a history of k-means algorithms", in BRITO P., CUCUMEL G., BERTRAND P., CARVALHO F. (eds.), *Selected Contributions in Data Analysis and Classification*, Springer-Verlag, Heidelberg, pp. 161–172, 2007.

[BOU 07] BOUVEYRON C., GIRARD S., SCHMID C., "High-dimensional data clustering", *Computational Statistics and Data Analysis*, vol. 52, no. 1, pp. 502–519, 2007.

[BRY 88] BRYANT P.G., "On characterizing optimization-based clustering criteria", *Journal of Classification*, vol. 5, pp. 81–84, 1988.

[BRY 05] BRYAN K., CUNNINGHAM P., BOLSHAKOVA N., *et al.*, "Biclustering of expression data using simulated annealing", *Proceedings of the 18th IEEE Symposium on Computer-Based Medical Systems*, pp. 383–388, 2005.

[CAI 76] CAILLIEZ F., PAGES J.P., *Introduction à l'analyse des données*, Smash, Paris, 1976.

[CAM 98] CAMIZ S., DENIMAL J., "A new method for cross-classification analysis of contingency data tables", in PAYNE R., GREEN P. (eds.), *COMPSTAT 98 - Proceedings in Computational Statistics*, Physica-Verlag, Heidelberg, pp. 209–214, 1998.

[CAR 84] CARAUX G., "Réorganisation et représentation visuelle d'une matrice de données numériques: un algorithme itératif", *Revue de Statistique Appliquée*, vol. 32, no. 4, pp. 5–23, 1984.

[CAR 00] CARREIRA-PERPIÑÁN M.Á., RENALS S., "Practical identifiability of finite mixtures of multivariate bernoulli distributions", *Neural Computation*, vol. 12, no. 1, pp. 141–152, 2000.

[CAR 06] CARMONA-SAEZ P., PASCUAL-MARQUI R.D., TIRADO F., *et al.*, "Biclustering of gene expression data by non-smooth non-negative matrix factorization", *BMC Bioinformatics*, vol. 7, no. 1, pp. 7–78, 2006.

[CEL 85] CELEUX G., DIEBOLT J., "Stochastic versions of the EM algorithm", *Computational Statistics Quarterly*, vol. 2, pp. 73–82, 1985.

[CEL 89] CELEUX G., DIDAY E., GOVAERT G., *et al.*, *Classification automatique des données: environnement statistique et informatique*, Dunod, Paris, 1989.

[CEL 91] CELEUX G., GOVAERT G., "Clustering criteria for discrete data and latent class models", *Journal of Classification*, vol. 8, no. 2, pp. 157–176, 1991.

[CEL 92] CELEUX G., GOVAERT G., "A classification EM algorithm for clustering and two stochastic versions", *Computational Statistics and Data Analysis*, vol. 14, no. 3, pp. 315–332, 1992.

[CEL 93] CELEUX G., GOVAERT G., "Comparison of the mixture and the classification maximum likelihood in cluster analysis", *Journal of Statistical Computation and Simulation*, vol. 47, pp. 127–146, 1993.

[CEL 95] CELEUX G., GOVAERT G., "Gaussian parsimonious clustering models", *Pattern Recognition*, vol. 28, no. 5, pp. 781–793, 1995.

[CEL 12] CELISSE A., DAUDIN J.J., PIERRE L., "Consistency of maximum-likelihood and variational estimators in the stochastic block model", *Electronic Journal of Statistics*, vol. 6, pp. 1847–1899, 2012.

[CHA 09] CHARRAD M., LECHEVALLIER Y., BEN A.M., *et al.*, "Block clustering for web pages categorization", *Intelligent Data Engineering and Automated Learning (IDEAL 09)*, Lecture Notes in Computer Science, vol. 5788, Burgos, Spain, pp. 260–267, September 2009.

[CHE 00] CHENG Y., CHURCH G., "Biclustering of expression data", *8th International Conference on Intelligent Systems for Molecular Biology (ISMB '00)*, San Diego, CA, pp. 93–103, 19–23 August 2000.

[CHO 04] CHO H., DHILLON I., GUAN Y., *et al.*, "Minimum sum-squared residue co-clustering of gene expression data", *Proceedings of the 4th SIAM International Conference on Data Mining*, pp. 114–125, April 2004.

[CHO 08] CHO H., DHILLON I.S., "Coclustering of human cancer microarrays using minimum sum-squared residue coclustering", *IEEE/ACM Transactions on Computational Biology and Bioinformatics*, vol. 5, no. 3, pp. 385–400, 2008.

[CIA 05] CIAMPI A., GONZALEZ M.A., CASTEJON L.M., "Correspondence analysis and two-way clustering", *Statistics and Operations Research Transactions*, vol. 29, no. 1, pp. 27–42, 2005.

[COO 50] COOMBS C.H., "Psychological scaling without a unit of measurement", *Psychological Review*, vol. 57, no. 3, pp. 145–158, 1950.

[COR 90] CORSTEN L., DENIS J.B., "Structuring interaction in two-way tables by clustering", *Biometrics*, vol. 46, no. 1, pp. 207–215, 1990.

[DAL 50] DALENIUS T., "The problem of optimum stratification", *Skandinavisk Aktuarietidskrift*, vol. 33, pp. 203–213, 1950.

[DAL 51] DALENIUS T., GURENEY M., "The problem of optimum stratification. II", *Skandinavisk Aktuarietidskrift*, vol. 34, pp. 133–148, 1951.

[DEM 77] DEMPSTER A.P., LAIRD N.M., RUBIN D.B., "Maximum likelihood from incomplete data via the EM algorithm", *Journal of the Royal Statistical Society. Series B (Methodological)*, vol. 39, no. 1, pp. 1–38, 1977.

[DEO 07] DEODHAR M., GHOSH J., "Simultaneous co-clustering and modeling of market data", *Data Mining for Marketing Workshop (ICDM'07)*, 2007.

[DER 96] DERISI J., PENLAND L., BROWN P.O., *et al.*, "Use of a cDNA microarray to analyse gene expression patterns in human cancer", *Nature Genetics*, vol. 14, no. 4, pp. 457–460, 1996.

[DHI 01] DHILLON I.S., "Co-clustering documents and words using bipartite spectral graph partitioning", *Proceedings of the 7th ACM SIGKDD International Conference on Knowledge Discovery and Data Mining (KDD '01)*, New York, NY, ACM, pp. 269–274, 2001.

[DHI 03] DHILLON I., MALLELA S., MODHA D., "Information-theoretic co-clustering", *Proceedings of the 9th ACM SIGKDD International Conference on Knowledge Discovery and Data Mining (KDD'03)*, pp. 89–98, 2003.

[DID 71] DIDAY E., "Une nouvelle méthode de classification automatique et reconnaissance des formes: la méthode des nuées dynamiques", *Revue de statistique appliquée*, vol. 19, no. 2, pp. 19–34, 1971.

[DID 74] DIDAY E., GOVAERT G., "Classification avec distances adaptatives", *Comptes Rendus de l'Académie des sciences Paris, série A*, vol. 278, pp. 993–995, 1974.

[DID 76] DIDAY E., SIMON J.C., "Clustering analysis", in FU K.S. (ed.), *Digital Pattern Recognition*, Springer-Verlag, Berlin, pp. 47–94, 1976.

[DID 77] DIDAY E., GOVAERT G., "Classification automatique avec distances adaptatives", *RAIRO Informatique / Computer Science*, vol. 11, no. 4, pp. 329–349, 1977.

[DID 79] DIDAY E., *Optimisation et classification automatique*, INRIA, Rocquencourt, France, 1979.

[DIJ 09] DIJK A.V., ROSMALEN J.V., PAAP R., A Bayesian approach to two-mode clustering, Econometric Institute Report no. EI 2009-06, Econometric Institute, Erasmus University, Rotterdam, 2009.

[DIN 06] DING C., LI T., PENG W., *et al.*, "Orthogonal nonnegative matrix t-factorizations for clustering", *Proceedings of the 12th ACM SIGKDD International Conference on Knowledge Discovery and Data Mining*, ACM, pp. 126–135, 2006.

[DUB 70] DUBIN R., CHAMPOUX J., Typology of empirical attributes: dissimilarity linkage analysis (DLA), Technical Report no. 3, University of California, 1970.

[DUF 91] DUFFY D.E., QUIROZ A.J., "A permutation-based algorithm for block clustering", *Journal of Classification*, vol. 8, pp. 65–91, 1991.

[ECK 93] ECKES T., ORLIK P., "An error variance approach to two-mode hierarchical clustering", *Journal of Classification*, vol. 10, no. 1, pp. 51–74, 1993.

[ERT 10] ERTEN C., SÖZDINLER M., "Improving performances of suboptimal greedy iterative biclustering heuristics via localization", *Bioinformatics*, vol. 26, no. 20, pp. 2594–2600, 2010.

[EVE 93] EVERITT B.S., *Cluster Analysis*, 3rd ed., Arnold, London, 1993.

[FIL 08] FILIPPONE M., CAMASTRA F., MASULLI F., *et al.*, "A survey of kernel and spectral methods for clustering", *Pattern Recognition*, vol. 41, no. 1, pp. 176–190, 2008.

[FIS 69] FISHER W., *Clustering and Aggregation in Economics*, Johns Hopkins Press, 1969.

[FOR 65] FORGY E.W., "Cluster analysis of multivariate data: efficiency versus interpretability of classification", *Biometrics*, vol. 21, no. 3, pp. 768–769, 1965.

[FRI 67] FRIEDMAN H.P., RUBIN J., "On some invariant criteria for grouping data", *Journal of American Statistical Association*, vol. 62, pp. 1159–1178, 1967.

[FRÜ 06] FRÜHWIRTH-SCHNATTER S., *Finite Mixture and Markov Switching Models*, Springer-Verlag, 2006.

[FRÜ 11] FRÜHWIRTH-SCHNATTER S., "Label switching under model uncertainty", *Mixtures: Estimation and Applications*, Wiley, pp. 193–218, 2011.

[GAR 86] GARCIA H., PROTH J.M., "A new cross-decomposition algorithm: the GPM comparison with the bond energy method", *Control and Cybernetics*, vol. 15, pp. 155–165, 1986.

[GEO 05] GEORGE T., "A scalable collaborative filtering framework based on co-clustering", *5th IEEE International Conference on Data Mining*, pp. 625–628, 2005.

[GIL 92] GILBERT J., MOLER C., SCHREIBER R., "Sparse matrices in MATLAB: design and implementation", *SIAM Journal on Matrix Analysis and Applications*, vol. 13, pp. 333–356, 1992.

[GOL 92] GOLDBERG D., NICHOLS D., OKI B., TERRY D., "Using collaborative filtering to weave an information tapestry", *Communications of the ACM*, vol. 35, no. 12, pp. 61–70, 1992.

[GOO 65] GOOD I., "Categorization of classification", *Mathematics and Computer Science in Medicine and Biology*, pp. 115–128, 1965.

[GOO 74] GOODMAN L.A., "Exploratory latent structure models using both identifiable and unidentifiable models", *Biometrika*, vol. 61, pp. 215–231, 1974.

[GOO 85] GOODMAN L., "The analysis of cross-classified data having ordered and/or unordered categories: association models, correlation models, and asymmetry models for contingency tables with or without missing entries", *The Annals of Statistics*, vol. 13, no. 1, pp. 10–69, 1985.

[GOR 02] GORDON G., JENSEN R.V., HSIAO L., *et al.*, "Translation of microarray data into clinically relevant cancer diagnostic tests using gene expression ratios in lung cancer and mesothelioma", *Cancer Research*, vol. 62, pp. 4963–4967, 2002.

[GOV 77] GOVAERT G., "Algorithme de classification d'un tableau de contingence", *1st International Symposium on Data Analysis and Informatics*, Versailles, INRIA, pp. 487–500, 1977.

[GOV 83] GOVAERT G., Classification croisée, Thesis, University of Paris 6, France, 1983.

[GOV 89] GOVAERT G., "Clustering model and metric with continuous data", in DIDAY E. (ed.), *Data Analysis, Learning Symbolic and Numeric Knowledge*, Nova Science Publishers, New York, pp. 95–102, 1989.

[GOV 90a] GOVAERT G., "Classification binaire et modèles", *Revue de Statistique Appliquée*, vol. XXXVIII, no. 1, pp. 67–81, 1990.

[GOV 90b] GOVAERT G., Modèle de classification et distance dans le cas discret, Report, UTC, Compiègne, France, 1990.

[GOV 95] GOVAERT G., "Simultaneous clustering of rows and columns", *Control and Cybernetics*, vol. 24, no. 4, pp. 437–458, 1995.

[GOV 96] GOVAERT G., NADIF N., "Comparison of the mixture and the classification maximum likelihood in cluster analysis when data are binary", *Computational Statistics and Data Analysis*, vol. 23, pp. 65–81, 1996.

[GOV 02] GOVAERT G., NADIF M., "Block Clustering on continuous data", *Workshop on clustering High Dimensional Data and Its Application, 2nd SIAM International Conference on Data Mining*, Arlington, pp. 7–16, 11–13 April 2002.

[GOV 03] GOVAERT G., NADIF M., "Clustering with block mixture models", *Pattern Recognition*, vol. 36, pp. 463–473, 2003.

[GOV 05] GOVAERT G., NADIF M., "An EM algorithm for the block mixture model", *IEEE Transactions on Pattern Analysis and Machine Intelligence*, vol. 27, no. 4, pp. 643–647, 2005.

[GOV 06] GOVAERT G., NADIF M., "Fuzzy clustering to estimate the parameters of block mixture models", *Soft Computing*, vol. 10, no. 5, pp. 415–422, 2006.

[GOV 07] GOVAERT G., NADIF M., "Clustering of contingency table and mixture model", *European Journal of Operational Research*, vol. 183, pp. 1055–1066, 2007.

[GOV 08] GOVAERT G., NADIF M., "Block clustering with Bernoulli mixture models: comparison of different approaches", *Computational Statistics and Data Analysis*, vol. 52, no. 6, pp. 3233–3245, 2008.

[GOV 10] GOVAERT G., NADIF M., "Latent block model for contingency table", *Communications in Statistics - Theory and Methods*, vol. 39, no. 3, pp. 416–425, 2010.

[GOW 74] GOWER J.C., "Maximal predictive classification", *Biometrics*, vol. 30, pp. 643–654, 1974.

[GRE 88a] GREENACRE M., "Clustering the rows and columns of a contingency table", *Journal of Classification*, vol. 5, pp. 39–51, 1988.

[GRE 88b] GREENACRE M., "Correspondence analysis of multivariate categorical data by weighted least-squares", *Biometrika*, vol. 75, no. 3, p. 457, 1988.

[GRE 07] GREENACRE M., *Correspondence Analysis in Practice*, Chapman & Hall/CRC, 2007.

[GUP 10] GUPTA N., AGGARWAL S., "MIB: using mutual information for biclustering gene expression data", *Pattern Recognition*, vol. 43, no. 8, pp. 2692–2697, 2010.

[GYL 94] GYLLENBERG M., KOSKI T., REILINK E., *et al.*, "Nonuniqueness in probabilistic numerical identification of bacteria", *Journal of Applied Probabilities*, vol. 31, no. 0, pp. 542–548, 1994.

[HAN 00] HANSOHM J., "Two-mode clustering with genetic algorithms", *24th Annual Conference of the Gesellschaft für Klassifikation*, pp. 87–93, 2000.

[HAN 11] HANCZAR B., NADIF M., "Using the bagging approach for biclustering of gene expression data", *Neurocomputing*, vol. 74, no. 10, pp. 1595–1605, 2011.

[HAN 12a] HAN J., KAMBER M., PEI J., *Data Mining: Concepts and Techniques, 3rd ed.*, Morgan Kaufmann, 2012.

[HAN 12b] HANCZAR B., NADIF M., "Ensemble methods for biclustering tasks", *Pattern Recognition*, vol. 45, no. 11, pp. 3938–3949, 2012.

[HAR 76] HARMAN H.H., *Modern Factor Analysis*, University of Chicago Press, Chicago, 1976.

[HAR 83] HARRIS R.R., KANJI G.K., "On the use of minimum chi-square estimation", *The Statistician, Journal of the Royal Statistical Society. Series D (The Statistician)*, vol. 32, no. 4, pp. 379–394, 1983.

[HAR 72] HARTIGAN J.A., "Direct clustering of a data matrix", *Journal of the American Statistical Association*, vol. 67, no. 337, pp. 123–129, 1972.

[HAR 75a] HARTIGAN J.A., *Clustering Algorithms*, Wiley, New York, 1975.

[HAR 75b] HARTIGAN J.A., "Printer Graphics for Clustering", *Journal of Statistical Computation and Simulation*, vol. 4, no. 3, pp. 187–213, 1975.

[HAR 00] HARTIGAN J.A., "Bloc voting in the united states senate", *Journal of Classification*, vol. 17, no. 1, pp. 29–49, 2000.

[HAT 86] HATHAWAY R.J., "Another interpretation of the EM algorithm for mixture distributions", *Statistics & Probability Letters*, vol. 4, pp. 53–56, 1986.

[HOF 75] HOFFER J.A., SEVERANCE D.G., "The use of cluster analysis in physical data base design", *Proceedings of the 1st International Conference on Very Large Data Bases (VLDB)*, ACM, New York,pp. 69–86, 1975.

[HOF 99a] HOFMANN T., PUZICHA J., JORDAN M., "Unsupervised learning from dyadic data", *Advances in Neural Information Processing Systems*, vol. 11, 1999.

[HOF 99b] HOFMANN T., PUZICHA J., "Latent class models for collaborative filtering", *Proceedings of the 16th International Joint Conference on Artificial Intelligence (IJCAI '99)*, Morgan Kaufmann Publishers Inc., San Francisco, CA, pp. 688–693, 1999.

[HUA 97] HUANG Z., "Clustering large data sets with mixed numeric and categorical values", *First Pacific-Asia Conference on Knowledge Discovery and Data Mining*, pp. 21–34, 1997.

[HUA 98] HUANG Z., "Extensions to the k-Means algorithm for clustering large data sets with categorical values", *Data Mining and Knowledge Discovery*, vol. 2, no. 3, pp. 283–304, 1998.

[IHM 02] IHMELS J., FRIEDLANDER G., BERGMANN S., *et al.*, "Revealing modular organization in the yeast transcriptional network", *Nature Genetics*, vol. 31, pp. 370–377, 2002.

[IHM 04a] IHMELS J., BERGMANN S., BARKAI N., "Defining transcription modules using large-scale gene expression data", *Bioinformatics*, vol. 20, no. 13, pp. 1993–2003, 2004.

[IHM 04b] IHMELS J., BERGMANN S., "Challenges and prospects in the analysis of large-scale gene expression data", *Briefings in Bioinformatics*, vol. 5, no. 4, pp. 313–327, 2004.

[JAG 07] JAGALUR M., PAL C., LEARNED-MILLER E., *et al.*, "Analyzing in situ gene expression in the mouse brain with image registration, feature extraction and block clustering", *BMC Bioinformatics*, vol. 8, (Suppl 10), p. S5, 2007.

[JAM 76] JAMBU M., "Sur l'interprétation mutuelle d'une classification hiérarchique et d'une analyse des correspondances", *Revue de Statistique Appliquée*, vol. 24, no. 2, pp. 45–73, 1976.

[JAN 66] JANCEY R.C., "Multidimensional group analysis", *Australian Journal of Botany*, vol. 14, no. 1, pp. 127–130, 1966.

[JIT 01] JITIAN X., YANCHUN Z., XIAOHUA J., "Clustering non-uniform-sized spatial objects to reduce I/O cost for spatial-join processing", *The Computer Journal*, vol. 44, no. 5, pp. 384–397, 2001.

[JOL 02] JOLLOIS F.X., NADIF M., "Clustering large categorical data", *Pacific-Asia Conference on Advances in Knowledge Discovery and Data Mining, PAKDD*, pp. 257–263, 2002.

[JOL 03] JOLLOIS F.X., Contribution de la classification automatique à la fouille de donneés, PhD Thesis, University of Metz, France, 2003.

[JOL 04] JOLLOIS F.X., NADIF M., "Identification of homogeneous blocks in large binary data sets", *Neural Networks and Computational Intelligence*, pp. 60–65, 2004.

[JON 72] JONES K., "A statistical interpretation of term specificity and its application in retrieval", *Journal of Documentation*, vol. 28, pp. 11–21, 1972.

[KAU 87] KAUFMAN L., ROUSSEEUW P., "Clustering by means of medoids", in DODGE Y. (ed.), *Statistical Data Analysis Based on the L1-Norm and Related Methods*, North-Holland Publishing Company, pp. 405–416, 1987.

[KAU 90] KAUFMAN L., ROUSSEEUW P., *Finding Groups in Data: An Introduction to Cluster Analysis*, Wiley, New York, 1990.

[KEM 06] KEMP C., TENENBAUM J., GRIFFITHS T., *et al.*, "Learning systems of concepts with an infinite relational model", *Proceedings of the National Conference on Artificial Intelligence*, 2006.

[KER 13] KERIBIN C., BRAULT V., CELEUX G., *et al.*, Estimation and selection for the latent block model on categorical data, Research Report no. RR-8264, INRIA, March 2013.

[KLE 08] KLEINBERG J., SANDLER M., "Using mixture models for collaborative filtering", *Journal of Computer and System Sciences*, vol. 74, no. 1, pp. 49–69, 2008.

[KOH 82] KOHONEN T., "Self organized formation of topological correct feature maps", *Biological Cybernetics*, vol. 43, pp. 59–69, 1982.

[LAB 11a] LABIOD L., NADIF M., "Co-clustering under nonnegative matrix tri-factorization", *ICONIP (2)*, ICONIP'11, pp. 709–717, 2011.

[LAB 11b] LABIOD L., NADIF M., "Co-clustering for binary and categorical data with maximum modularity", *2011 IEEE 11th International Conference on Data Mining (ICDM)*, IEEE, pp. 1140–1145, 2011.

[LAS 09] LASHKARI D., GOLLAND P., Co-clustering with generative models, Report, MIT, 2009.

[LAZ 68] LAZARFIELD P.F., HENRY N.W., *Latent Structure Analysis*, Houghton Mifflin Company, Boston, 1968.

[LAZ 00] LAZZERONI L., OWEN A., Plaid models for gene expression data, Report, Stanford University, 2000.

[LEB 84] LEBART L., MORINEAU A., WARWICK K., *Multivariate Descriptive Statistical Analysis*, Wiley, New York, 1984.

[LER 77] LERMAN I.C., LEREDDE H., "La méthodes des pôles d'attraction", *1st International Symposium on Data Analysis and Informatics*, Versailles, North Holland, 1977.

[LER 80] LEREDDE H., PERIN P., "Les plaques-boucles mérovingiennes", *Les dossiers de l'Archéologie*, vol. 42, pp. 83–87, 1980.

[LER 81] LERMAN I.C., *Classification automatique et analyse ordinale des données*, Dunod, Paris, 1981.

[LI 05] LI T., "A general model for clustering binary data", *Proceedings of the 11th ACM SIGKDD International Conference on Knowledge Discovery in Data Mining*, ACM, pp. 188–197, 2005.

[LI 06] LI J., ZHA H., "Two-way Poisson mixture models for simultaneous document classification and word clustering", *Computational Statistics & Data Analysis*, vol. 50, no. 1, pp. 163–180, 2006.

[LIN 80] LINDE Y., BUZO A., GRAY R., "An algorithm for vector quantizer design", *IEEE Transactions on Communications*, vol. 28, no. 1, pp. 84–95, 1980.

[LIU 03] LIU J., WANG W., "OP-Cluster: clustering by tendency in high dimensional space", *3rd IEEE International Conference on Data Mining*, pp. 187–194, 2003.

[LLO 57] LLOYD S.P., Least squares quantization in PCM's, Report, Bell Telephone Laboratories Paper, Murray Hill, 1957.

[LON 05] LONG B., ZHANG Z., YU P., "Co-clustering by block value decomposition", *Proceedings of the 11th ACM SIGKDD International Conference on Knowledge Discovery in Data Mining*, New York, NY, ACM, pp. 635–640, 2005.

[MAC 67] MACQUEEN J.B., "Some methods for classification and analysis of cluster analysis", in LECAM L.M., NEYMAN J. (eds.), *Proceedings of 5th Berkeley Symposium on Mathematics, Statistics and Probability*, vol. 1, University of California Press, pp. 281–297, 1967.

[MAD 04] MADEIRA S.C., OLIVEIRA A.L., "Biclustering algorithms for biological data analysis: a survey", *IEEE/ACM Transactions on Computational Biology and Bioinformatics*, vol. 1, no. 1, pp. 24–45, 2004.

[MAR 72] MAROY J.P., PENEAU J.P., Analyse des données et conception en architecture, Technical Report, no. 13, IRIA, Le Chesnay, 1972.

[MAR 75] MARRIOTT F.H.C., "Separating mixtures of normal distributions", *Biometrics*, vol. 31, pp. 767–769, 1975.

[MAR 87] MARCOTORCHINO F., "Block seriation problems: a unified approach", *Applied Stochastic Models and Data Analysis*, vol. 3, pp. 73–91, 1987.

[MAR 91a] MARCOTORCHINO F., "Seriation problems: an overview", *Applied Stochastic Models and Data Analysis*, vol. 7, no. 2, pp. 139–151, 1991.

[MAR 91b] MARTINETZ T., SCHULTEN K., "A "neural-gas" network learns topologies", in KOHONEN T., MAKISARA K., SIMULA O., KANGAS J. (eds.), *Artificial Neural Networks*, North-Holland, Publishing Company, Amsterdam, pp. 397–402, 1991.

[MCC 72] MCCORMICK W., SCHWEITZER P., WHITE T., "Problem decomposition and data reorganization by a clustering technique", *Operations Research,*, vol. 20, no. 5, pp. 993–1009, 1972.

[MCL 82] MCLACHLAN G.J., "The classification and mixture maximum likelihood approaches to cluster analysis", in KRISHNAIAH P.R., KANAL L.N. (eds.), *Handbook of Statistics*, vol. 2, North-Holland Publishing Company, pp. 199–208, 1982.

[MCL 88] MCLACHLAN G.J., BASFORD K.E., *Mixture Models, Inference and Applications to Clustering*, Marcel Dekker, New York, 1988.

[MCL 97] MCLACHLAN G.J., KRISHNAN K., *The EM Algorithm*, Wiley, New York, 1997.

[MCL 00] MCLACHLAN G.J., PEEL D., *Finite Mixture Models*, Wiley, New York, 2000.

[MCL 07] MCLACHLAN G.J., KRISHNAN T., *The EM Algorithm and Extensions*, vol. 382, Wiley-Interscience, 2007.

[MEE 07] MEEDS E., ROWEIS S., Nonparametric Bayesian biclustering, Technical Report no. UTML-TR-2007-001, University of Toronto, June 2007.

[MIT 06] MITRA S., BANKA H., "Multi-objective evolutionary biclustering of gene expression data", *Pattern Recognition*, vol. 39, 2006.

[NAD 93] NADIF M., MARCHETTI F., "Classification de données qualitatives et modèles", *Revue de statistique appliquée*, vol. 41, no. 1, pp. 55–69, 1993.

[NAD 10] NADIF M., GOVAERT G., "Model-based co-clustering for continuous data", *International Conference on Machine Learning and Applications, ICMLA*, pp. 175–180, 2010.

[NAV 84] NAVATHE S., CERI S., WIEDERHOLD G., *et al.*, "Vertical partitioning algorithms for database design", *ACM Transactions on Database Systems*, vol. 9, no. 4, pp. 680–710, 1984.

[NEA 98] NEAL R.M., HINTON G.E., "A view of the EM algorithm that justifies incremental, sparse, and other variants", in JORDAN M.I. (ed.), *Learning in Graphical Models*, Kluwer Academic Publishers, Dordrecht, pp. 355–358, 1998.

[NEY 49] NEYMAN J., "Contribution to the theory of x2 test", *Proceedings of the Berkeley Symposium on Mathematical Statistics and Probability*, Berkeley, University of California Press, pp. 239–73, 1949.

[NG 94] NG R., HAN J., "Efficient and effective clustering methods for spatial data mining", in BOCCA J.B., JARKE M., ZANIOLO C. (eds.), *Proceedings of 20th International Conference on Very Large Data Bases (VLDB'94)*, Santiago de Chile, Chile, Morgan Kaufmann, pp. 144–155, 12–15 September 1994.

[NIE 05] NIERMANN S., "Optimizing the ordering of tables with evolutionary computation", *Journal of the American Statistical Association*, vol. 59, no. 1, pp. 41–46, 2005.

[NOW 01] NOWICKI K., SNIJDERS T.A.B., "Estimation and prediction for stochastic blockstructures", *Journal of the American Statistical Association*, vol. 96, no. 455, pp. 1077–1087, 2001.

[OYA 01] OYANAGI S., KUBOTA K., NAKASE A., "Application of matrix clustering to web log analysis and access prediction", *WEBKDD 2001 Mining Web Log Data Across All Customers Touch Points, 3rd International Workshop*, pp. 13–21, 2001.

[PEA 00] PEARSON K., "On the criterion that a given system of deviations from the probable in the case of a correlated system of variables is such that it can be reasonably supposed to have arisen from random sampling", *The London, Edinburgh, and Dublin Philosophical Magazine and Journal of Science*, vol. 50, no. 302, pp. 157–175, 1900.

[PEN 05] PENSA R.G., ROBARDET C., BOULICAUT J.F., "A biclustering framework for categorical data", *Proceedings of the 9th European Conferences on Principles and Practice of Knowledge Discovery in Databases (PKDD)*, Lecture Notes in Computer Science, vol. 3721, Springer-Verlag, pp. 643–650, 2005.

[PEN 08] PENG W., LI T., SHAO B., "Clustering multi-way data via adaptive subspace iteration", *Proceeding of the 17th ACM Conference on Information and Knowledge Management (CIKM '08)*, New York, NY, ACM, pp. 1519–1520, 2008.

[POD 91] PODANI J., FEOLI E., "A general strategy for the simultaneous classification of variables and objects in ecological data tables", *Journal of Vegetation Science*, vol. 2, no. 4, pp. 435–444, 1991.

[POI 08] POIRIER D., BOTHOREL C., BOULLÉ M., "Analyse exploratoire d'opinions cinématographiques: co-clustering de corpus textuels communautaires", *Extraction et gestion des connaissances (EGC'08)*, pp. 565–576, 2008.

[POL 02] POLLARD K.S., VAN DER LAAN M., "Statistical inference for simultaneous clustering of gene expression data", *Mathematical Biosciences*, vol. 176, no. 1, pp. 99–121, 2002.

[PRE 06] PRELIC A., BLEULER S., ZIMMERMANN P., *et al.*, "A systematic comparison and evaluation of biclustering methods for gene expression data", *Bioinformatics*, vol. 22, no. 9, pp. 1122–1129, 2006.

[PRI 12] PRIAM R., NADIF M., GOVAERT G., "Nonlinear mapping by constrained co-clustering", *International Conference on Pattern Recognition Applications and Methods, ICPRAM (1)*, vol. 1, pp. 63–68, 2012.

[PRI 13] PRIAM R., NADIF M., GOVAERT G., "Gaussian topographic co-clustering model", *International Symposium on Intelligent Data Analysis, IDA*, pp. 345–356, 2013.

[PUO 08] PUOLAMÄKI K., HANHIJÄRVI S., GARRIGA G.C., "An approximation ratio for biclustering", *Information Processing Letters*, vol. 108, no. 2, pp. 45–49, 2008.

[ROC 08] ROCCI R., VICHI M., "Two-mode multi-partitioning", *Computational Statistics & Data Analysis*, vol. 52, no. 4, pp. 1984–2003, 2008.

[ROO 95] ROOTH M., "Two-dimensional clusters in grammatical relations", in S.U. (ed.), *Symposium on Representation and Acquisition of Lexical Knowledge: Polysemy, Ambiguity, and Generativity*, AAAI Spring Symposium Series, Stanford University, 1995.

[ROS 09] VAN ROSMALEN J., GROENEN P.J., *et al.*, "Optimization strategies for two-mode partitioning", *Journal of Classification*, vol. 26, no. 2, pp. 155–181, 2009.

[ROU 11] ROUSSEAU J., MERGENSEN K., "Asymptotic behaviour of the posterior distribution in overfitted models", *Journal of the Royal Statistical Society*, vol. 73, pp. 689–710, 2011.

[RUS 69] RUSPINI E.H., "A new approach to clustering", *Information and Control*, vol. 15, pp. 22–32, 1969.

[SAL 83] SALTON G., MCGILL M., *Introduction to Modern Information Retrieval*, McGraw-Hill, New York, 1983.

[SCH 76] SCHROEDER A., "Analyse d'un mélange de distributions de probabilité de même type", *Revue de Statistique Appliquée*, vol. 24, no. 1, pp. 39–62, 1976.

[SCH 77a] SCHROEDER A., "Étude de traces de références de programmes: analyse de processus à régimes", *1st International Symposium on Data Analysis and Informatics*, Rocquencourt, INRIA, 1977.

[SCH 77b] SCHROEDER A., A statistical approach to the study of program behavior via reference strings analysis, IRIA, Technical Report, June, 1977.

[SCH 78] SCHWARZ G., "Estimating the number of components in a finite mixture model", *The Annals of Statistics*, vol. 6, no. 2, pp. 461–464, 1978.

[SCH 02] SCHÖLKOPF B., SMOLA A.J., *Learning with Kernels: Support Vector Machines, Regularization, Optimization, and Beyond*, MIT Press, London, 2002.

[SCH 83] SCHROEDER A., Une étude quantitative statique de programmes Pascal, Report no. RT-0029, INRIA, Rocquencourt, France, 1983.

[SCO 71] SCOTT A.J., SYMONS M.J., "Clustering methods based on likelihood ratio criteria", *Biometrics*, vol. 27, pp. 387–397, 1971.

[SHA 06] SHAFIEI M.,, MILIOS E., "Model-based overlapping co-clustering", *Proceedings of the 4th Workshop on Text Mining, 6th SIAM International Conference on Data Mining*, Bethesda, Maryland, April 2006.

[SHA 08] SHAN H., BANERJEE A., "Bayesian co-clustering", *Proceedings of the 2008 Eighth IEEE International Conference on Data Mining (ICDM '08)*, Washington, DC, IEEE Computer Society, pp. 530–539, 2008.

[SIB 73] SIBSON R., "SLINK: an optimally efficient algorithm for the single-link cluster method", *The Computer Journal*, vol. 16, no. 1, pp. 30–45, 1973.

[SLO 00] SLONIM N., TISHBY N., Y Y.I., "Document clustering using word clusters via the information bottleneck method", *International ACM SIGIR conference on Research and development in information retrieval, ACM SIGIR 2000*, ACM, pp. 208–215, 2000.

[SOK 58] SOKAL R.R., MICHENER C.D., "A statistical method for evaluating systematic relationships", *The University of Kansas Science Bulletin*, vol. 38, pp. 1409–1438, 1958.

[SOR 48] SORENSEN T., "A method of establishing groups of equal amplitude in plant sociology based on similarity of species content and its applications to analyses of the vegetation on Danish commons", *Bioligiske Skrifter*, vol. 5, pp. 1–34, 1948.

[STE 56] STEINHAUSS H., "Sur la division des corps matériels en parties", *Bulletin de l'Académie Polonaise des Sciences*, Class III, vol. 4, no. 12, pp. 801–804, 1956.

[STO 51] STOUFFER S.A., TOBY J., "Role conflict and personality", *American Journal of Sociology*, vol. 56, pp. 395–506, 1951.

[SYM 81] SYMONS M.J., "Clustering criteria and multivariate normal mixture", *Biometrics*, vol. 37, pp. 35–43, March 1981.

[TAK 02] TAKAMURA H., MATSUMOTO Y., "Two-dimensional clustering for text categorization", *Proceedings of the 6th Conference on Natural Language Learning (COLING'02)*, Morristown, NJ, Association for Computational Linguistics, pp. 1–7, 2002.

[TAN 02] TANAY A., SHARAN R., SHAMIR R., "Discovering statistically significant biclusters in gene expression data", *Bioinformatics*, vol. 18 (Suppl 1), pp. 136–144, 2002.

[TAN 05] TANAY A., SHARAN R., SHAMIR R., "Biclustering algorithms: a survey", *Chapman & Hall/CRC Computer and Information*, vol. 9, pp. 1–20, 2005.

[TCH 06] TCHAGANG A.B., TEWFIK A.H., "DNA microarray data analysis: a novel biclustering algorithm approach", *EURASIP Journal on Applied Signal Processing*, vol. 2006, pp. 1–12, 2006.

[TIB 99] TIBSHIRANI R., HASTIE T., EISEN M., *et al.*, Clustering methods for the analysis of DNA microarray data, Report, Department of Statistics, Stanford University, 1999.

[TIS 99] TISHBY N., PEREIRA F.C., BIALEK W., "The information bottleneck method", *The 37th Annual Allerton Conference on Communication, Control, and Computing*, pp. 368–377, 1999.

[TIT 85] TITTERINGTON D.M., SMITH A.F.M., MAKOV U.E., *Statistical Analysis of Finite Mixture Distributions*, Wiley, New York, 1985.

[TOL 77] TOLEDANO J., BROUSSE J., "Une méthode de classification simultanée des lignes et des colonnes d'un tableau", *Premières Journées Internationales Analyse des Données et Informatique*, IRIA, pp. 105–107, 1977.

[TRY 70] TRYON R., BAILEY D., *Cluster Analysis*, McGraw-Hill, New York, 1970.

[TUC 64] TUCKER L.R., "The extension of factor analysis to three-dimensional matrices", in GULLIKSEN H., FREDERIKSEN N. (eds.), *Contributions to Mathematical Psychology*, Holt, Rinehart and Winston, New York, pp. 110–127, 1964.

[TUR 05] TURNER H., BAILEY T., KRZANOWSKI W., "Improved biclustering of microarray data demonstrated through systematic performance tests", *Computational Statistics & data Analysis*, vol. 48, no. 2, pp. 235–254, 2005.

[UNG 98] UNGAR L., FOSTER D., "Clustering methods for collaborative filtering", *AAAI Workshop on Recommendation Systems*, pp. 112–125, 1998.

[VAN 04] VAN MECHELEN I., BOCK H.H., DE BOECK P., "Two-mode clustering methods: a structured overview", *Statistical Methods in Medical Research*, vol. 13, no. 5, pp. 363–394, 2004.

[VIC 01] VICHI M., "Double k-means clustering for simultaneous classification of objects and variables", *Advances in Classification and Data Analysis*, Springer, pp. 43–52, 2001.

[WIN 87] WINDHAM M.P., "Parameter modification for clustering criteria", *Journal of Classification*, vol. 4, pp. 191–214, 1987.

[WON 82] WONG M.A., "A hybrid clustering method for identifying high-density clusters", *Journal of the American Statistical Association*, vol. 77, no. 380, pp. 841–847, 1982.

[XU 10] XU G., ZONG Y., DOLOG P., *et al.*, "Co-clustering analysis of weblogs using bipartite spectral projection approach", *Proceedings of the 14th International Conference on Knowledge-Based and Intelligent Information and Engineering Systems: Part III*, KES'10, Berlin, Heidelberg, Springer-Verlag, pp. 398–407, 2010.

[YOO 10] YOO J., CHOI S., "Orthogonal nonnegative matrix tri-factorization for co-clustering: multiplicative updates on Stiefel manifolds", *Information Processing and Management*, vol. 46, no. 5, pp. 559–570, 2010.

[ZHA 10] ZHANG Z., LI T., DING C., *et al.*, "Binary matrix factorization for analyzing gene expression data", *Data Mining and Knowledge Discovery*, vol. 20, no. 1, pp. 28–52, 2010.

# Index

www.ingramcontent.com/pod-product-compliance
Ingram Content Group UK Ltd.
Pitfield, Milton Keynes, MK11 3LW, UK
UKHW020214290926
15470UKWH00001B/174

9 781848 214736